Biological Materials and Structures in Physiologically Extreme Conditions and Disease

MATERIALS RESEARCH SOCIETY
SYMPOSIUM PROCEEDINGS VOLUME 1274

Biological Materials and Structures in Physiologically Extreme Conditions and Disease

Spring 2010, April 5–9, San Francisco, California, U.S.A.

EDITORS:

Markus J. Buehler

Massachusetts Institute of Technology
Cambridge, Massachusetts, U.S.A.

David Kaplan

Tufts University
Medford, Massachusetts, U.S.A.

Chwee Teck Lim

National University of Singapore
Singapore, Singapore

Joachim Spatz

University of Heidelberg
Stuttgart, Germany

Materials Research Society
Warrendale, Pennsylvania

CAMBRIDGE UNIVERSITY PRESS
Cambridge, New York, Melbourne, Madrid, Cape Town,
Singapore, São Paulo, Delhi, Mexico City

Cambridge University Press
32 Avenue of the Americas, New York NY 10013-2473, USA

Published in the United States of America by Cambridge University Press, New York

www.cambridge.org
Information on this title: www.cambridge.org/9781107406742

Materials Research Society
506 Keystone Drive, Warrendale, PA 15086
http://www.mrs.org

First published 2010
First paperback edition 2012

Single article reprints from this publication are available through
University Microfilms Inc., 300 North Zeeb Road, Ann Arbor, MI 48106

CODEN: MRSPDH

ISBN 978-1-605-11251-0 Hardback
ISBN 978-1-107-40674-2 Paperback

CONTENTS

Preface ix

Materials Research Society Symposium Proceedings.................... x

CELL MECHANICS AND CELL-MATERIAL INTERACTIONS I

* The Role of Mechanical Tension in Neurons.......................... 3
  Jagannathan Rajagopalan, Alireza Tofangchi,
  and M. Taher A. Saif

CELL MECHANICS AND CELL-MATERIAL INTERACTIONS II

Modeling and Simulation of Soft Contact and Adhesion of
Stem Cells... 11
  Shaofan Li and Xiaowei Zeng

Analyzing the Mesoscopic Structure of Pericellular Coats on
Living Cells... 19
  H. Boehm, T.A. Mundinger, V. Hagel,
  C.H.J. Boehm, J.E. Curtis, and J.P. Spatz

* The Contribution of the N- and C- Terminal Domains to the
  Stretching Properties of Intermediate Filaments 25
  Laurent Kreplak

ECTOPIC MATERIALS IN THE CONTEXT OF PRION AND AMYLOID DISEASES

* Quantitative Approaches for Characterising Fibrillar Protein
  Nanostructures.. 33
  Duncan A. White, Christopher M. Dobson,
  Mark E. Welland, and Tuomas P.J. Knowles

*Invited Paper

POSTER SESSION

Controlling the Physical Properties of Random Network Based Shape Memory Polymer Foams 43
Pooja Singhal, Thomas S. Wilson,
and Duncan J. Maitland

Crystallization of Synthetic Hemozoin (BetaHematin) Nucleated at the Surface of Synthetic Neutral Lipid Bodies 51
Anh N. Hoang, Kanyile K. Ncokazi,
Katherine A. de Villiers, David W. Wright,
and Timothy J. Egan

The Effect of Type 1 Diabetes on the Structure and Function of Fibrillin Microfibrils 57
R. Akhtar, J.K. Cruickshank, N.J. Gardiner,
B. Derby, and M.J. Sherratt

NANOMECHANICS OF BIOLOGICAL MOLECULES AND STRUCTURES

A Strategy to Simulate the Dynamics of Molecular Assemblies Over Long Times 65
Julius T. Su

Mechanics of Collagen in the Human Bone: Role of Collagen-Hydroxyapatite Interactions. 71
Shashindra M. Pradhan, Kalpana S. Katti,
and Dinesh R. Katti

Humidity Induced Softening Leads to Apparent Capillary Effect in Gecko Adhesion 81
Bin Chen and Huajian Gao

STRUCTURAL MATERIALS AND TISSUES I

Computational Multiscale Studies of Collagen Tissues in the Context of Brittle Bone Disease *osteogenesis imperfecta* 89
Alfonso Gautieri, Simone Vesentini,
Alberto Redaelli, and Markus J. Buehler

Effect of Environment on Mechanical Properties of Micron-Sized Beams of Bone Fabricated Using FIB 95
Ines Jimenez-Palomar and Asa H. Barber

STRUCTURAL MATERIALS AND TISSUES II

* The Scissors Model of Microcrack Detection in Bone: Work in
Progress 103
     David Taylor, Lauren Mulcahy, Gerardo Presbitero,
     Pietro Tisbo, Clodagh Dooley, Garry Duffy,
     and T. Clive Lee

Mechanical Testing of Limpet Teeth Micro-Beams
Using FIB 113
     Dun Lu and Asa H. Barber

Experimental Investigations into the Mechanical Properties of
the Collagen Fibril-Noncollagenous Protein (NCP) Interface in
Antler Bone 119
     Fei Hang and Asa H. Barber

In Vitro Study of a Novel Method to Repair Human Enamel 125
     Song Yun, Yanjun Ge, Yujing Yin, Hailan Feng,
     and Haifeng Chen

MULTISCALE SOFT TISSUE MECHANICS

Surgical Adhesive/Soft Tissue Adhesion Measured by
Pressurized Blister Test...................................... 133
     Muriel L. Braccini, Bertrand R.M. Perrin,
     Cécile Bidan, and Michel Dupeux

Author Index 139

Subject Index 141

*Invited Paper

PREFACE

Symposium QQ, "Biological Materials and Structures in Physiologically Extreme Conditions and Disease," held April 5–9 at the 2010 MRS Spring Meeting in San Francisco, California, brought together researchers who work at the interface of materials science and biology, with a focus on biological materials under extreme physical, chemical as well as physiological conditions, and human disease. The symposium focused on the integration of advanced experimental, computational and theoretical methods, utilized to assess structure-process-property relations and to monitor and predict mechanisms associated with failure of protein materials and structures composed of them. A focal point of the studies presented was the study of materials phenomena that play a crucial role in disease etiology, progression and treatment, covering multiple scales, and ranging from nano to macro.

The study of biological materials under extreme conditions includes changes in forces, pressures, pH, chemical environments and other factors, and how materials fail to provide the designed biological functions as a result of these altered conditions. Advances in experimental and computational methods make it now possible to develop a systematic understanding of complex materials phenomena across all scales, and to integrate the physical sciences, biology and medicine. The presentations included studies of materials failure in the context of genetic diseases such as *osteogenesis imperfecta* (brittle bone disease), rapid aging disease *progeria*, as well as other medical disorders such as Alzheimer's disease, infectious disease (e.g. malaria), as well as cancer. The study of the role of materials in disease and materials failure mechanisms represents an innovative approach that opens exciting new avenues for materials research with broad impact in the life sciences, bio- and nanomedicine, the development of new biologically inspired materials, and beyond.

We hope that readers will greatly enjoy reading the papers included in this book, and that it may stimulate much further research in this field of study. We thank all authors for the careful preparation of papers for inclusion in the proceedings volume.

We greatly acknowledge symposium support from the Air Force Office of Scientific Research and the Army Research Office. This support is much appreciated.

Markus J. Buehler
David Kaplan
Chwee Teck Lim
Joachim Spatz

October 2010

MATERIALS RESEARCH SOCIETY SYMPOSIUM PROCEEDINGS

Volume 1245 — Amorphous and Polycrystalline Thin-Film Silicon Science and Technology—2010, Q. Wang, B. Yan, C.C. Tsai, S. Higashi, A. Flewitt, 2010, ISBN 978-1-60511-222-0

Volume 1246 — Silicon Carbide 2010—Materials, Processing and Devices, S.E. Saddow, E.K. Sanchez, F. Zhao, M. Dudley, 2010, ISBN 978-1-60511-223-7

Volume 1247E —Solution Processing of Inorganic and Hybrid Materials for Electronics and Photonics, 2010, ISBN 978-1-60511-224-4

Volume 1248E —Plasmonic Materials and Metamaterials, J.A. Dionne, L.A. Sweatlock, G. Shvets, L.P. Lee, 2010, ISBN 978-1-60511-225-1

Volume 1249 — Advanced Interconnects and Chemical Mechanical Planarization for Micro- and Nanoelectronics, J.W. Bartha, C.L. Borst, D. DeNardis, H. Kim, A. Naeemi, A. Nelson, S.S. Papa Rao, H.W. Ro, D. Toma, 2010, ISBN 978-1-60511-226-8

Volume 1250 — Materials and Physics for Nonvolatile Memories II, C. Bonafos, Y. Fujisaki, P. Dimitrakis, E. Tokumitsu, 2010, ISBN 978-1-60511-227-5

Volume 1251E —Phase-Change Materials for Memory and Reconfigurable Electronics Applications, P. Fons, K. Campbell, B. Cheong, S. Raoux, M. Wuttig, 2010, ISBN 978-1-60511-228-2

Volume 1252— Materials and Devices for End-of-Roadmap and Beyond CMOS Scaling, A.C. Kummel, P. Majhi, I. Thayne, H. Watanabe, S. Ramanathan, S. Guha, J. Mannhart, 2010, ISBN 978-1-60511-229-9

Volume 1253 — Functional Materials and Nanostructures for Chemical and Biochemical Sensing, E. Comini, P. Gouma, G. Malliaras, L. Torsi, 2010, ISBN 978-1-60511-230-5

Volume 1254E —Recent Advances and New Discoveries in High-Temperature Superconductivity, S.H. Wee, V. Selvamanickam, Q. Jia, H. Hosono, H-H. Wen, 2010, ISBN 978-1-60511-231-2

Volume 1255E —Structure-Function Relations at Perovskite Surfaces and Interfaces, A.P. Baddorf, U. Diebold, D. Hesse, A. Rappe, N. Shibata, 2010, ISBN 978-1-60511-232-9

Volume 1256E —Functional Oxide Nanostructures and Heterostructures, 2010, ISBN 978-1-60511-233-6

Volume 1257 — Multifunctional Nanoparticle Systems—Coupled Behavior and Applications, Y. Bao, A.M. Dattelbaum, J.B. Tracy, Y. Yin, 2010, ISBN 978-1-60511-234-3

Volume 1258 — Low-Dimensional Functional Nanostructures—Fabrication, Characterization and Applications, H. Riel, W. Lee, M. Zacharias, M. McAlpine, T. Mayer, H. Fan, M. Knez, S. Wong, 2010, ISBN 978-1-60511-235-0

Volume 1259E —Graphene Materials and Devices, M. Chhowalla, 2010, ISBN 978-1-60511-236-7

Volume 1260 — Photovoltaics and Optoelectronics from Nanoparticles, M. Winterer, W.L. Gladfelter, D.R. Gamelin, S. Oda, 2010, ISBN 978-1-60511-237-4

Volume 1261E —Scanning Probe Microscopy—Frontiers in NanoBio Science, C. Durkan, 2010, ISBN 978-1-60511-238-1

Volume 1262 — In-Situ and Operando Probing of Energy Materials at Multiscale Down to Single Atomic Column—The Power of X-Rays, Neutrons and Electron Microscopy, C.M. Wang, N. de Jonge, R.E. Dunin-Borkowski, A. Braun, J-H. Guo, H. Schober, R.E. Winans, 2010, ISBN 978-1-60511-239-8

Volume 1263E —Computational Approaches to Materials for Energy, K. Kim, M. van Shilfgaarde, V. Ozolins, G. Ceder, V. Tomar, 2010, ISBN 978-1-60511-240-4

Volume 1264 — Basic Actinide Science and Materials for Nuclear Applications, J.K. Gibson, S.K. McCall, E.D. Bauer, L. Soderholm, T. Fanghaenel, R. Devanathan, A. Misra, C. Trautmann, B.D. Wirth, 2010, ISBN 978-1-60511-241-1

Volume 1265 — Scientific Basis for Nuclear Waste Management XXXIV, K.L. Smith, S. Kroeker, B. Uberuaga, K.R. Whittle, 2010, ISBN 978-1-60511-242-8

Volume 1266E —Solid-State Batteries, S-H. Lee, A. Hayashi, N. Dudney, K. Takada, 2010, ISBN 978-1-60511-243-5

Volume 1267 — Thermoelectric Materials 2010—Growth, Properties, Novel Characterization Methods and Applications, H.L. Tuller, J.D. Baniecki, G.J. Snyder, J.A. Malen, 2010, ISBN 978-1-60511-244-2

Volume 1268 — Defects in Inorganic Photovoltaic Materials, D. Friedman, M. Stavola, W. Walukiewicz, S. Zhang, 2010, ISBN 978-1-60511-245-9

Volume 1269E —Polymer Materials and Membranes for Energy Devices, A.M. Herring, J.B. Kerr, S.J. Hamrock, T.A. Zawodzinski, 2010, ISBN 978-1-60511-246-6

MATERIALS RESEARCH SOCIETY SYMPOSIUM PROCEEDINGS

Volume 1270 — Organic Photovoltaics and Related Electronics—From Excitons to Devices,
V.R. Bommisetty, N.S. Sariciftci, K. Narayan, G. Rumbles, P. Peumans, J. van de Lagemaat,
G. Dennler, S.E. Shaheen, 2010, ISBN 978-1-60511-247-3

Volume 1271E —Stretchable Electronics and Conformal Biointerfaces, S.P. Lacour, S. Bauer, J. Rogers,
B. Morrison, 2010, ISBN 978-1-60511-248-0

Volume 1272 — Integrated Miniaturized Materials—From Self-Assembly to Device Integration,
C.J. Martinez, J. Cabral, A. Fernandez-Nieves, S. Grego, A. Goyal, Q. Lin, J.J. Urban,
J.J. Watkins, A. Saiani, R. Callens, J.H. Collier, A. Donald, W. Murphy, D.H. Gracias,
B.A. Grzybowski, P.W.K. Rothemund, O.G. Schmidt, R.R. Naik, P.B. Messersmith,
M.M. Stevens, R.V. Ulijn, 2010, ISBN 978-1-60511-249-7

Volume 1273E —Evaporative Self Assembly of Polymers, Nanoparticles and DNA , B.A. Korgel, 2010,
ISBN 978-1-60511-250-3

Volume 1274 — Biological Materials and Structures in Physiologically Extreme Conditions and Disease,
M.J. Buehler, D. Kaplan, C.T. Lim, J. Spatz, 2010, ISBN 978-1-60511-251-0

Prior Materials Research Society Symposium Proceedings available by contacting Materials Research Society

Cell Mechanics and Cell-Material Interactions I

Mater. Res. Soc. Symp. Proc. Vol. 1274 © 2010 Materials Research Society   1274-QQ01-06

The Role of Mechanical Tension in Neurons

Jagannathan Rajagopalan, Alireza Tofangchi, M. Taher A. Saif
Department of Mechanical Science and Engineering,
University of Illinois at Urbana-Champaign

ABSTRACT

We used high resolution micromechanical force sensors to study the *in vivo* mechanical response of embryonic Drosophila neurons. Our experiments show that Drosophila axons have a rest tension of a few nN and respond to mechanical forces in a manner characteristic of viscoelastic solids. In response to fast externally applied stretch they show a linear force-deformation response and when the applied stretch is held constant the force in the axons relaxes to a steady state value over time. More importantly, when the tension in the axons is suddenly reduced by releasing the external force the neurons actively restore the tension, sometimes close to their resting value. Along with the recent findings of Siechen et al (Proc. Natl. Acad. Sci. USA **106**, 12611 (2009)) showing a link between mechanical tension and synaptic plasticity, our observation of active tension regulation in neurons suggest an important role for mechanical forces in the functioning of neurons in vivo.

INTRODUCTION

The influence of mechanical forces/microenvironment on various cell processes such as motility, growth and differentiation has become increasingly clear over the last two decades [1, 2]. In particular, neurons have been shown to respond to a variety of mechanical inputs. For example, in vitro experiments on different neuronal cells have revealed that new axons can be initiated by externally applied tension [3, 4] and that existing axons grow when tension that exceeds a threshold is applied [5, 6]. Experiments have also shown that a buildup of mechanical tension in a developing axonal branch stabilizes it and causes the retraction of other axon branches and collaterals [7].

A recent experimental study by Siechen et al [8] on live Drosophila embryos has revealed an important new facet of the role of mechanical tension in neurons. These experiments show that mechanical tension is necessary for accumulation of neurotransmitter vesicles in the pre-synaptic terminal of Drosophila motor neurons. Furthermore, an increase in tension enhances the accumulation of the vesicles. This connection between tension and vesicle accumulation suggests that neurons are likely to regulate their tension in vivo. To test this hypothesis, we studied the *in vivo* mechanical response of Drosophila axons using high resolution micromechanical force sensors. Our experiments reveal the overall mechanical behavior of axons and provide direct evidence that neurons regulate their tension in response to mechanical perturbations.

EXPERIMENTAL DETAILS

Transgenic *Drosophila (ELAV-GAP/ GFP)* embryos expressing green fluorescent protein (GFP) in neuronal membranes were used for the experiments. Flies were transferred from fly stock into a culture container sealed with grape jelly-coated petri dish and maintained under controlled temperature (25 C) and humidity (~60%). After 20-24 hours in the culture container, eggs hatched on the petri dish were washed with 50% diluted bleach, filtered and rinsed with deionized (DI) water for 2 minutes. Under fluorescent light, eggs at correct age were separated and placed on a double sided tape affixed to a glass cover slide for dissection. The cover slide was functionalized with *3-Amino propyltrie thoxysilane* (APTES) solution to improve the adherence of the embryos to the surface during dissection. Under saline solution, egg membranes were pierced using glass micro needles and the embryos were extracted and placed on the cover slide. Embryos were oriented such that the ventral nerve cord was closest to the glass surface. Dorsal incisions were made and the embryos were flattened out against the glass. The gut and fat cells around the axon of interest were gently scraped off using the needle until a single axon was exposed and isolated.

The isolated axons were deformed using micromechanical force sensors and their force response was monitored simultaneously. The fabrication and working of the force sensors have been described elsewhere [9]. The movement of the force sensors was controlled using an x-y-z piezo actuator. Live imaging of the axon under the applied deformation was carried out using an inverted Olympus microscope. The time-lapse images were analyzed using ImageJ to measure the deformation and force on the individual axons.

The mechanical properties of the axons were investigated by studying their response to systematic stretching experiments, which comprised of the following steps:

1. In the first step, the axons were loaded within a period of 1-2 minutes to a predetermined level of stretch (usually less than 40% of the axon length) using the force sensor.
2. The force sensor was then held fixed and the time evolution of axonal force was recorded over a period of 10-15 minutes.
3. After force relaxation to a steady state, the force sensor was quickly (within 1-2 minutes) backed up to release the force on the axon.

RESULTS AND DISCUSSIONS

During fast loading (step 1) a linear relationship between axonal force and applied deformation was found in all the axons. In effect, the axons behave like elastic springs when subject to sudden changes in force (Fig. 1). The stiffness of the axons, given by the slope of the force-deformation curve, varied from embryo to embryo with values ranging from 5 nN/μm to 50 nN/μm. To verify whether axons maintain a rest tension, we extrapolated the force deformation curve during loading to zero deformation. The extrapolation yielded a positive force value for each of the axons, confirming that the axons maintain a rest tension in vivo.

When the force sensor was held fixed after loading, the force in the axons decreased over time. An initial fast decay in force was followed by a more gradual decrease to a steady state value over a period of 10-15 minutes in all the experiments. The steady state value of the axonal force was typically larger when the force applied during loading was larger. Figure 2 shows the force relaxation in a typical axon over time.

4

When the force sensor was backed up to reduce the force on the axon in step 3, the force-deformation response of the axon was again linear but the slope of the curve showed some variation from loading. In some of the cases, the stiffness of the axon during unloading was higher than the loading stiffness where as in other cases it was lower. There was also variability in the response of the axons after the external force was completely removed. Several axons immediately regained their taut appearance after unloading. On the other hand, some of the axons had a slack appearance after unloading, indicating a lack of tension. However, these axons actively contracted over time and in the process built up the tension. The tension buildup in one of the axons after unloading is shown in Fig. 3. As evident from the figure, the force build up mirrors the force relaxation in the axon - an initial fast increase is followed a more gradual approach to a steady state over time. The similarity between the time evolution of force relaxation and force build up suggests that the same biological mechanism may underlie both phenomena. More importantly, this force buildup in the axons after unloading cannot be accounted by a standard viscoelastic model, that is, this restoration of tension in the axons represents an active biological response of the neurons. In other words, neurons actively regulate the tension in the axons in vivo.

As mentioned earlier, recent experiments [8] have shown that tension is necessary for the accumulation of neurotransmitter vesicles at the pre-synaptic terminal of motor neurons. When the axon is severed using laser axotomy, thereby releasing the tension, the density of vesicles reduces more than threefold compared to intact axons. However, when tension is resupplied by pulling the severed end with a micropipette the vesicle accumulation increases significantly. Furthermore, increasing the tension in an axon by applying an external force increases the vesicle accumulation to more than two times that of unstretched axons, which shows a direct relationship between mechanical tension and clustering of neurotransmitter vesicles. Our experiments provide further support, albeit indirectly, to the notion that tension plays a vital role in neuronal function by demonstrating that neurons actively regulate the tension in the axons in vivo.

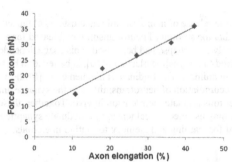

Figure 1: Force-deformation response of a typical axon to fast stretching. Extrapolation of the curve to zero axon elongation results in a positive force, which is the rest tension in the axon.

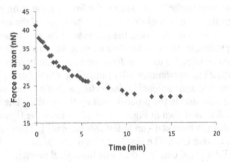

Figure 2: Force relaxation in an axon over time. An initial fast decrease in force is followed by a more gradual reduction to a steady state value over a period of 10-15 min.

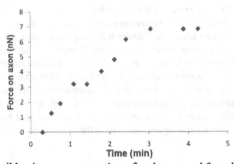

Figure 3: Force buildup in an axon over time after the external force has been released.

It is well known that the release of neurotransmitters clustered at the synapse is essential for neurotransmission and that the amount of neurotransmitters released for a given action potential depends on how frequently the synapse had been used. This usage dependent release of neurotransmitters, referred to as synaptic plasticity [10], is believed to be the basis for learning and memory formation in animals. The findings of Siechen et al. [8] show that mechanical tension influences the accumulation of neurotransmitters in the synapse and our experiments strongly suggest that neurons regulate their tension in vivo. Together, these results raise the possibility that neurons may use mechanical tension to modulate synaptic plasticity and that tension may be essential for learning and memory formation in animals.

CONCLUSIONS

The mechanical behavior of neurons in live Drosophila embryos was studied using high resolution micromechanical force sensors. Drosophila axons behave like viscoelastic solids in response to externally applied force but, in addition, also exhibit active force generation to

6

restore tension after mechanical perturbation. These results show that neurons regulate the tension in the axons and suggest that tension may be critical for neuronal function *in vivo*.

ACKNOWLEDGEMENTS

This work was supported by National Institutes of Health grant NIH/NINDS NS063405-01 and the National Science Foundation grants NSF ECS 05-24675, NSF CMMI 0800870, NSF ECCS 0801928. The force sensors were fabricated in the Micro and Nanomechanical Systems Cleanroom at the University of Illinois at Urbana-Champaign.

REFERENCES

1. R. J. Pelham Jr. and Y. Wang, *Proc. Natl. Acad.Sci. USA* **94**, 13661 (1997).
2. T. Yeung, P. C. Georges, L. A. Flanagan, B. Marg, M. Ortiz, M. Funaki, N. Zahir, W. Ming, V. Weaver, and P. A. Janmey, *Cell Motility and Cytoskeleton* **60**, 24 (2005).
3. J. Zheng, P. Lamoureux, V. Santiago, T. Dennerll, R. E. Buxbaum, and S. R. Heidemann, *J. Neurosci.* **11**, 1117 (1991).
4. S. Chada, P. Lamoureux, R. E. Buxbaum, and S. R. Heidemann, *J. Cell Sci.* **110**, 1179 (1997).
5. D. Bray. Axonal growth in response to experimentally applied mechanical tension. *Dev. Bio.* **102**, 379 (1984).
6. T. J. Dennerll, P. Lamoureux, R. E. Buxbaum, and S. R. Heidemann, *J. Cell Bio.* **109**, 3073 (1989).
7. S. Anava, A. Greenbaum, E. B. Jacob, Y. Hanein, and A. Ayali, *Biophys. J.* **96**, 1661 (2009).
8. S. Siechen, S. Yang, A. Chiba, and T. Saif, *Proc. Natl. Acad. Sci. USA* **106**, 12611 (2009).
9. J. Rajagopalan, A. Tofangchi and M. T. A. Saif, *Proc. IEEE 23rd Intl Conf. on MEMS*, pp. 88-91, doi 10.1109/MEMSYS.2010.5442558 (2010).
10. E.R. Kandel, J. H. Schwartz, T. M. Jessell, in *Principles of Neural Science* (McGraw Hill, New York, 2000).

Cell Mechanics and Cell-Material Interactions II

Mater. Res. Soc. Symp. Proc. Vol. 1274 © 2010 Materials Research Society   1274-QQ02-02

Modeling and Simulation of Soft Contact and Adhesion of Stem Cells

Shaofan Li and Xiaowei Zeng
Department of Civil and Environmental Engineering,
University of California, Berkeley, CA94720, U.S.A.

ABSTRACT

In this paper, we briefly report our recent work on multiscale modeling and simulations of soft elasticity and focal adhesion of stem cells. In particular, our work is focused on modeling and simulation of contact and adhesion of stem cells on substrates with different rigidities. In order to understand the precise mechanical influences on cell contact/adhesion and to explain the possible mechanotransduction mechanism, we have developed a three-dimensional soft-matter cell model that uses liquid-crystal gel or liquid-crystal elastomer gel to model the overall constitutive relations of the cell, and we have simulated the responses of the cell to extra-cellular stimulus. The discussion here is specifically focused on the following issues: (1) how to model the overall myosin responses at the early stage of differentiation process of the stem cell, (2) the effects of both the adhesive force due to ligand-receptor interaction or focal adhesion and the surface tension, and (3) possible cell conformation and configuration changes triggered by substrate's rigidity.

INTRODUCTION

Recently, experimental results have shown that the rigidity and micro-structures of the substrate have significant influences on the differentiation of stem cells or the fate of stem cells. For example Discher et al [3] and Engler et al [4] reported that matrix elasticity directs stem cell lineage specification. Rehfeldt et al [5] reported that results with drug treatments of various cells on soft, stiff, and rigid matrices show a broad range of possible matrix-dependent drug responses; and cells on soft gels might be relatively unaffected in cell spreading or apoptosis-induction whereas cells on stiff substrates seem more sensitive to diverse drugs in terms of spreading. All these indicate a significant influence of matrix elasticity on cell contact or adhesion, and subsequent cytoskeleton reorganization. To explain such biophysical or physiology phenomenon has been one of the current focuses of stem cell research as well as the cell mechanics and biophysics research.

Because of its scientific and clinic importance, a major research focus of molecular cell biology is the study of mechanotransduction of cells, in particular stem cells. The physical process of mechanotransduction of cells is through contact and adhesion between cells and their extra-cellular environment. Recently, several mechanics models of cell contact and focal adhesion have been proposed, for example Freund & Lin [9] and Deshpande et. al. [10]. To further advance bio-physics modeling of stem cells, in this work, we have build a three-dimensional cell model by treating cells or stem cells as a special soft matter. A multi-component three-dimensional cell model with a coarse grain adhesion body force and various surface tension expressions has been formulated and constructed. In our model, the utmost layer is a layer of nematic or smetic-A liquid crystal, and the inner cell is modeled as a hyperelastic core with or without liquid crystal elastomer contents. The extra-cellular matrix is modeled as a substrate of hyperelastic block and liquid-crystal polymer layer with or without liquid crystal elastomer

11

content. The extra-cellular matrix is modeled as a substrate of hyperelastic block or liquid-crystal polymer layer.

CELL AND CELL CONTACT MODEL

In this work, we systematically build a three dimensional soft matter cell model by treating cells as soft matters. A multi-component three-dimensional cell model with a coarse grain adhesion body force and various surface tension expressions are proposed.

Basic hypothesis and assumptions

The cell membrane is basically a lipid bilayer. Up to today, the most successful cell model is the fluid mosaic model — the lipid bilayer model [1]. This model captures two essential features of the lipid bilayer: (1) fluidity and (2) diffusion. A well-established and very successful mechanics or mathematics model for the cell membrane is Helfrich's liquid crystal cell membrane model [2], which is based on or built on the fluid mosaic model. Because Helfrich's liquid crystal cell membrane model has successfully predicted the bi-concave shape of red blood cells, it has been regarded as the first triumph of soft matter physics. The success of this model is largely because the microstructure of the liquid crystal, especially that of Smectic-A liquid crystals, resembles that of the lipid bilayer of cells. In fact, many view the lipid bilayer as a form of bio-liquid-crystal. However, three decades after the Singer-Nicolson model, new evidences have shown that the freedom of proteins on cell membrane are far from unrestricted. Damjanovich [8] showed that the emerging evidence on hierarchically built super-structured protein complexes, which may hinder the diffusion of proteins in the membrane. Jacobson et al. [9] have pointed out: *"Most membrane proteins do not enjoy the continuous unrestricted lateral diffusion.... Instead, proteins diffuse in a more complicated way that indicates considerable lateral heterogeneity in membrane structure, at least on a nanometer scale."* This makes us consider a possible liquid crystal elastomer model to model the cell out layer as a weakly cross-linked network. Three-dimensional cell models are scarce, and the current state of mechanics study of cell contact and adhesion is elementary. From structure viewpoint, a cell consists of membrane wall, cytoplasm, microtubes, cell nucleus, and cytoskeleton — cell's scaffold. However, for stem cells, the cell cytoskeleton may be less developed than that of other functional cells. Since the cytoskeleton of stem cells are still in developing stage, we may first model the cell as a liquid crystal thin layer containing either liquid crystal elastomer or a hyperelastic core. This may provide a good approximation. Based on the recent research in cellular mechanotransduction, the mechanical forces that exerted on surface-adhesion receptors are channeled along actin filaments, and its molecular mechanism largely depends on the interaction between actin and myosin. If one exams the microstructure of myosin II in a cell, one may find that the mesogen of a Nematic liquid crystal elastomer may correspond to the myosin head. Considering the fact that the microstructure of myosin II is very similar to that of Nematic liquid crystal elastomer, we propose to use a nematic liquid crystal elastomer model to model the aggregates of actomyosin filaments.

We are making the following assumptions:
1. To capture the receptor diffusion and fluidity interactions between the protein and lipid, the outer layer of the cell may be modeled as a layer of liquid crystal or liquid crystal elastomer; 2. Considering that stem cells have a less developed cell scaffold structure, we model the

12

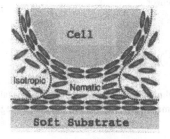

Figure 1. Soft matter cell model and contact model.

mechanical content of stem cells as either a hyperelastic material or liquid crystal elastomer; 3. To simulate mechanical responses of actomyosins, we model the actomyosin aggregates as a Nematic liquid crystal elastomer.

Liquid crystal model for cells

We adopt a simplified version of the Ericksen-Leslie theory (Lin and Liu [10]) as the governing equations for the nematic liquid crystal used in the cell modeling. The strong form of the simplified Ericksen-Leslie theory is as follows,

$$\rho_f \frac{D\mathbf{v}}{Dt} = \nabla \cdot \boldsymbol{\sigma} + \mathbf{b}, \quad \forall \mathbf{x} \in V(t), \qquad (1)$$

$$\rho_d \frac{D\mathbf{h}}{Dt} = \gamma(\nabla \cdot \nabla \otimes \mathbf{h} - \mathbf{f}(\mathbf{h})), \forall \mathbf{x} \in V(t), (2)$$

where $\mathbf{v}$ is the velocity field, $\boldsymbol{\sigma}$ is the stress field, ρ_f is the density of fluid, ρ_d is the density for the direct field, $\mathbf{h}$ is the Nematic director distribution field, γ is the Nematic elasticity constant, and $\mathbf{f}$ ($\mathbf{h}$) is the force generated by a Landau-Ginzburg type of potential,

$$\mathbf{f}(\mathbf{h}) = \frac{dR(\mathbf{h})}{d\mathbf{h}} \frac{1}{\epsilon^2}(|\mathbf{h}|^2 - 1), \text{ and } R(\mathbf{h}) = \frac{1}{4\epsilon^2}(|\mathbf{h}|^2 - 1)^2, \qquad (3)$$

and the Cauchy stress can be found as

$$\boldsymbol{\sigma} = -p\mathbf{I} + 2\mu\mathbf{d} - \lambda(\nabla \otimes \mathbf{h} \odot \nabla \otimes \mathbf{h}) \qquad (4)$$

In Eq. (4), p is the hydrostatic pressure, μ is the viscosity, λ is a positive constant. In Eq. (2), $\frac{D\mathbf{h}}{Dt}$ is denoted as the objective rate, and it can be chosen as the following rates:

1. The convected rate : $\overset{\circ}{\mathbf{h}} = \frac{D\mathbf{h}}{Dt} + \mathbf{l}^T \mathbf{h};$

2. The corotational rate : $\overset{\triangle}{\mathbf{h}} = \frac{D\mathbf{h}}{Dt} - \mathbf{wh};$

$$(5)$$

13

where **l** is the velocity gradient, and **w** is the spin tensor

$$\mathbf{l} = \dot{\mathbf{F}}\mathbf{F}^{-1} \qquad (6)$$

$$\mathbf{w} = \frac{1}{2}(\mathbf{l} - \mathbf{l}^T) \qquad (7)$$

We use the Mooney-Rivlin model to model the cell nucleus, and use either liquid crystal or liquid crystal elastomer to model the cell actin filament network, and cell plasma aggregate.

Liquid crystal elastomer model
A typical entropic free-energy expression for liquid crystal elastomer is

$$\mathcal{F}_{bulk} = \frac{1}{2}k_B T \cdot tr\left(\mathcal{L}_0 \cdot \mathbf{F}^T \cdot \mathcal{L}^{-1} \cdot \mathbf{F}\right) + \frac{1}{2}k_B T \ln\left(\frac{Det(\mathcal{L})}{a^3}\right) \qquad (8)$$

where k_B is the Boltzmann constant, T is temperature, tr is the trace operator, $\mathcal{L}_0$ and $\mathcal{L}$ are polymer's step length tensors in the referential and current configurations respectively. They are related to the director field as follows

$$\mathcal{L} = \ell_\perp \mathbf{I}^{(2)} + (\ell_\parallel - \ell_\perp)\mathbf{h} \otimes \mathbf{h}, \qquad (9)$$

We can then define

$$s = \frac{\ell_\parallel}{\ell_\perp}, \qquad (10)$$

Based on the free-energy expression, we can find the first Piola-Kirchhoff stress tensor by taking derivative of the free-energy with respect to deformation gradient,

$$\mathbf{P} = \frac{\partial \mathcal{F}}{\partial \mathbf{F}}, \qquad (11)$$

However, this free-energy potential is not convex, which poses a serious challenge in numerical computation of the Galerkin weak formulation. To regularize the free-energy potential, we add the Oseen-Zocher-Frank energy density—a strain gradient term to the bulk potential,

$$\mathcal{F}_{grad} = J\frac{\kappa(s-1)^2}{2s}|\mathbf{F}^{-T}\nabla_X \otimes \mathbf{h}|^2 = J\frac{\kappa(s-1)^2}{2s}|\mathbf{F}^{-T}\mathbf{G}|^2 \qquad (12)$$

where κ is a simplified Oseen-Zocher-Frank elastic constant, J is the determinant of Jacobian, and $\mathbf{G} = \nabla \otimes \mathbf{h}$. Then the total potential energy of the Nematic liquid crystal elastomor becomes,

$$\mathcal{F}_t = \frac{\mu}{2}(|\mathbf{F}|^2 - \frac{s-1}{s}|\mathbf{F}^T\mathbf{h}|^2 + (s-1)|\mathbf{F}\mathbf{h_0}|^2$$
$$- \frac{(s-1)^2}{s}(\mathbf{F}^T\mathbf{h}\cdot\mathbf{h_0})^2 - 3) + \frac{\kappa(s-1)^2}{2s}|\mathbf{F}^{-T}\nabla\otimes\mathbf{h}|^2 \quad (13)$$

The first dynamic equation of motion for the liquid crystal elastomer can be derived from the balance of linear momentum,

$$\rho_f\frac{\partial^2\mathbf{u}}{\partial t^2} = -\mathbf{F}^{-T}\nabla p + \text{DIV}\frac{\partial\mathcal{F}_t}{\partial\mathbf{F}} + \mathbf{B}, \quad (14)$$

where p is the hydrostatic pressure, Div is the divergence operator, and $\mathbf{B}$ is the body force. Even though a few people have tried to develop the dynamic equation for the director field of liquid crystal elastomers, how to establish thermodynamically correct hydrodynamics equations for the liquid crystal elastomers under finite deformations is still an open problem in soft matter physics. Considering the fact that we are more interested in the evolution of the order parameter, which may correlate the physical phenomena such as receptor diffusion or myosin head rotatory relaxation, we propose to adopt the following dynamic equation for direct field based on an Allen-Cahn type of approach [11],

$$\frac{D\tilde{\mathbf{h}}}{Dt} = -c\frac{\delta\Psi_t}{\delta\mathbf{h}} = \frac{\kappa(s-1)^2}{s}\nabla\cdot\left\{\mathbf{F}^{-1}J\mathbf{C}\mathbf{F}^{-T}\mathbf{G}\right\}$$
$$+ \mu c\left\{\frac{(s-1)}{s}(\mathbf{F}^T\mathbf{h})\mathbf{F}^T + \frac{(s-1)^2}{s}(\mathbf{F}^T\mathbf{h}\cdot\mathbf{h^0})(\mathbf{F}^T\cdot\mathbf{h^0})\right\} \quad (15)$$

A numerical model of the liquid crystal/liquid crystal elastomer cell is displayed in Fig. 1(b). A comprehensive exposition on modeling and computational algorithm of the soft matter cell model will be presented in several full-length papers elsewhere.

SIMULATION RESULTS

As Discher et al pointed out: "a major challenge in all such modeling is to clarify the principal enigma: how contractile traction forces exerted by a cell tend to increase with stiffness of the cell's substrate." [3]. Simulations have been carried out for cells in contact with four substrates of different elastic moduli or coarse-grained adhesive potentials. Our simulations have shown some remarkable features of cell contact and adhesion that have never been observed before, such as (1) cell spreading induced by combination of adhesive force and surface tension and its sensitivity to substrate rigidities; (2) cell shape evolution and the evolution of the mesogen director orientation distribution, an indicator of overall myosin structure of the stem cell, and (3) evolutions of the traction force between the cell and substrate and the overall contractile force inside the cell.

In the first example, the cell is modeled as a liquid crystal spherical drop. In Fig. 2, we display a time sequence of a cell contacting with a deformable substrate. The color contour is the effective stress contour. One may observe the cell spreading over time. Since both the cell and the substrate deform at the same time, the effective stress contours are displayed on both bodies.

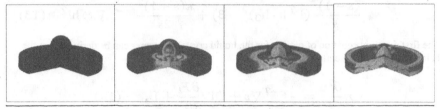

Figure 2. Time sequence of cell contact

In the second example, the cell is modeled as an aggregate of hyper-elastic core (nucleus) and liquid crystal elastomer outer layer. Fig 3 displays different cell shapes that are corresponding to on three substrates of different stiffness under the same contact condition at the same time. One may find that the contact between the cell and rigid substrate (Fig. 3 (a)) generates the most cell spreading, and the contact between the cell and a stiff elastic substrate (Fig. 3(b)) has the second most cell spreading, and the cell spreading for the cell on the softest substrate (Fig. 3(c)) is the least. This indicates that the cell spreading due to contact is directly related to the stiffness of the substrate, and it may be purely a phenomenon of soft elasticity.

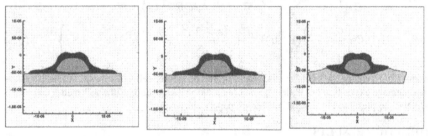

Figure 3. Cell spreading on substrates with different stiffness: (a) Rigid substrate, (b) Stiff but deformable substrate, and (c) Soft substrate.

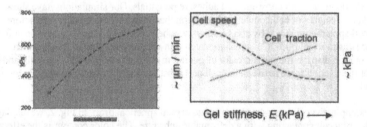

Figure 4. Comparison of cell traction on the substrate: (a) Simulation, and (b) From [3]

16

In Fig. 4, we plot the magnitude of the normal traction force versus values of substrate stiffness (Fig. 4(a)), and it can be clearly observed that as the substrate stiffness increases the magnitude of the traction force increases as well. This is qualitatively in agreement with the experimental observations reported by Discher et al [3] (Fig. 4 (b)).

ACKNOWLEDGMENTS

This research is supported by A. Richard Newton Research Breakthrough Award from Microsoft Corporation and a grant from National Science Foundation (CMMI No. 0800744).

REFERENCES

[1] Singer, S. J., and Nicolson, G. L. [1972] "Tissue cells feel and respond to the stiffness of their substrate". *Science*, **310**, 1134-1139.
[2] Helfrich, W. [1972],"Elastic properties of lipid bilayer: theory and possible experiments". *Z. Naturforsch. C*, **28**. 693-703.
[3] Discher, D. E., Janmey, P., and Wang, Y. L. [2005],"Tissue cells feel and respond to the stiffness of their substrate". *Science*, **175**. 720-731.
[4] Engler, A. J., Sen, S., Sweeney, H. L., and Discher, D. E. [2006]"Matrix elasticity directs stem cell lineage specification". *Cell*, **126**, 677-689.
[5] F. Rehfelt, A. E., Eckhart, A., Ahmed, F., and Discher, D. E. [2007],"Cell responses to the mechanochemical microenvironment—implications for regenerative medicine and drug delivery". *Advanced Drug Delivery Reviews*, **59**, 1329-1339.
[6] Freund, L. B., and Lin, Y. [2004] "The role of binder mobility in spontaneous adhesive contact and implications for cell adhesion". *Journal of the Mechanics and Physics of Solids*, **52**,2455-2472.
[7] Deshpande, V. S., Mrksich, M., McMeeking, R. M., and Evans, A. G. [2008]"A bio-mechanical model for coupling cell contractility with focal adhesion formation". *Journal of the Mechanics and Physics of Solids*, **56**, 1484-1510.
[8] Damjanovich, R. Gaspar, J., and Pieri, C. [1997],"Dynamic receptor superstructures at the plasma membrane". *Quarterly Reviews of Biophysics*, **30**, 67-106.
[9] Jacobson, K., Sheets, E., and Simon, R. "Revisiting the fluid mosaic model of membranes". *Science*, **268**, 1441-1442.
[10] Lin, F. H., and Liu, C. "Existence of solutions for the Ericksen-Leslie system". *Arch. Rat. Mech. Anal.*, **154**, 135-156.
[11] Cahn, J. W., and Allen, S. M. [1977]"A microscopic theory of domain wall motion and its experimental verification in feal alloy domain growth kinetics". Journal of Physics, C. **38**,7-51.

Mater. Res. Soc. Symp. Proc. Vol. 1274 © 2010 Materials Research Society   1274-QQ02-03

Analyzing the Mesoscopic Structure of Pericellular Coats on Living Cells

H. Boehm[1], T. A. Mundinger[1], V. Hagel[1], C. H. J. Boehm[1], J. E. Curtis[1,2], J. P. Spatz[1]

[1]Max-Planck-Institute for Metals Research, Department New Materials & Biosystems, Heisenbergstr. 3, 70569 Stuttgart, Germany & University of Heidelberg, Department of Biophysical Chemistry, INF 253, 69120 Heidelberg, Germany.
[2]School of Physics, Georgia Institute of Technology, 837 State Street, Atlanta, GA, USA

ABSTRACT

We employed passive particle-tracking microrheology to map the micromechanical structure of the hyaluronan-rich pericellular coat enveloping chondrocytes. Therefor we exploited the technique's position sensitivity to gain radial information on the coat. We observed a linear increase in viscoelasticity from the coat's rim towards the cell membrane. This gradient corresponds to hyaluronan concentration profiles observed in confocal fluorescent microscopy with small, specific hyaluronan markers. These results suggest that the structural basis of the pericellular coat is formed by grafted hyaluronan of different effective lengths stretched out by a homogenous decoration with hyaladherins such as aggrecan. The different effective lengths could be caused either by different lengths of the HA chains or by "side-on" attachments within the chain. Remarkably, the hyaluronan-rich coat increases the viscosity of the pericellular space only by about a factor of about two at 100 and at 20 Hz compared to pure media and an increasing elastic component is observed. Both the viscoelasticity as well as the hyaluronan concentration decrease linearly or slightly exponential from the cell membrane towards the PCC's rim. These observations could be obtained on living cells exploiting this unintrusive measurement techniques.

INTRODUCTION

Hyaluronan, a linear muco-polysaccharide forms the vital backbone of the highly hydrated pericellular coats (PCC) enveloping a variety of mammalian cells.[1] The hyaluronan (HA) chains of several micrometer lengths[2,3,4] serve as polyvalent templates binding several different proteins, called hyaladherins.[5,2] Some of these, such as CD44 or RHAMM, are membrane proteins grafting the hyaluronan to the cell surface.[3] Others like aggrecan are required for the formation of the PCC[6] and might serve to stiffen the hyaluronan even at very low concentrations.[7] The interplay of HA's polymer properties and its structural organization and interaction with hyaladherins influence its micromechanical features and thus the adhesion process,[8] proliferation, motility and embryogenesis[9].

First averaged elasticity values could be obtained with strain apparati[10,11] and micropipet aspiration[12] and could even distinguish between arthritic and normal human chondrocytes[13]. Local measurements at different areas of the cell can be performed with atomic force microscopy.[14,15,16] However, so far all of the applied rheometric techniques can not gather information about the micromechanical structure of the PCC, which determines the PCC's interaction with biomolecules and its mutability during proliferation and migration. The structure of the PCC will also influence the cell's perception of the extracellular environment including cell-cell contacts. Thus determining the PCC's structure is critical in understanding the mechanism behind biological functions of the PCC.

Here we employ a combination of a fluorescent and a rheological technique on the pericellular coat of living RCJ-P cells,[17] which serve as a well established model system for hyaluronan-rich coats[18,19]. Initially, we characterized the distribution of hyaluronan with confocal fluorescent microscopy and observed a linear decrease of fluorescent intensity towards the edge of the PCC. Additionally, we used passive particle tracking microrheology,[20,21] which is based on the detection of the thermal motion of added tracer particles. The velocity of the particle's Brownian motion is determined by the viscoelastic properties of the surrounding solution. Thus the position sensitive analysis of the acquired traces leads to a map of the micromechanical properties of the PCC without perturbing the cell or its coat. Corresponding to the fluorescent results, the elastic component as well as the viscosity increase linearly towards the rim of the PCC.

EXPERIMENT

Hyaluronan profiles in the PCC. Our investigation of the hyaluronan distribution within the pericellular coat (PCC) is based on its specific fluorescent labeling. As marker we attached the neurocan-GFP fusion protein (GFPn) developed by Zhang et al.[22] It is composed of an eGFP labeled neurocan link module which specifically recognizes HA.[22] Additionally, we measured the extension of the PCC indirectly to rule out structural changes induced by GFPn addition: in a particle exclusion assay (PEA) in which erythrocytes are excluded from the cell and its coat, but sediment densely around the uncovered volume (Fig. 1). The thickness of the PCC, as determined by the PEA, did not change upon GFPn addition. As expected, no staining could be observed, if the cells were treated with hyaluronidase first. This enzyme digests hyaluronan and thus removes the complete PCC, without affecting the cells otherwise. A more detailed description of the materials and methods is available in our Soft Matter paper.[17]

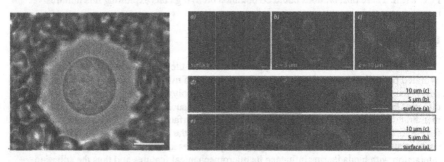

Fig. 1 (left): The PCC is clearly visible in a particle exclusion assay: In this indirect visualization method red blood cells are excluded from the volume taken up by the PCC, which is otherwise not discernible in any bright field microscopy technique due to its extremely high water content. The scale bar correspond to 20 μm.

Fig. 2 (right): RCJ-P cells are surrounded by their PCC on all non-adhered membranes: Confocal spinning disk images of GFPn stained RCJ-P cells after 24 h. Images of adhered cells were taken at different heights above the surface. The reconstituted images along the blue dotted line (d) and the dashed orange line (e) were performed on 25 images. Scale bars correspond to 10 μm.

Confocal spinning disc microscopy confirms the presence of the PCC also on the apical side of the cell (Fig. 2): the GFPn labeling leads to a comparable staining on all non-adhered sides and the three dimensional PEA employing fluorecent red blood cells reveals PCC thicknesses of about 3 to 8 μm around the cells.

We examined the distribution of hyaluronan within the PCC 4 μm above the surface which corresponds to about half the total height of an adhered cell (Fig. 3a). The fluorescent intensity is colocalized with the cell membrane and decreases towards the rim of the PCC (Fig. 3c). Intensity profiles chosen perpendicular to the cell membrane confirm this very reproducible curve progression (Fig. 3d).

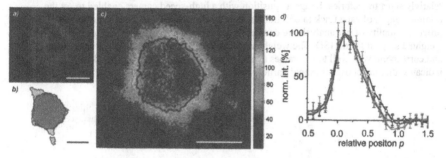

Fig. 3: Profiles of hyaluronan concentration within the PCC: a) Confocal spinning disk images of GFPn stained HA and WGA stained membranes (red) are taken 4 μm above the surface. b) The outline of the cell is deduced from the membrane staining at the surface (light red) and at 4 μm height. c) The false color image of the GFPn intensity shows the decreasing HA concentration towards the rim, which is d) highly reproducible when several cells are compared, here the relative distributions for three cells is shown. Scale bars correspond to 10 μm.

The decrease in intensity could either be described by an exponential ($e^{-x/const.}$) or a linear function. This concentration gradient is independent of membrane curvature, i.e. we observed it both on straight and curved membrane sections.[17] Due to the high noise ratio, even after summing over 10 pixels, reliable exponential fits are hard to perform. In order to compare the different intensity profiles and corresponding to the largely linear decrease, we fitted a linear slope corresponding to $I = m \cdot d + I(0)$, where I is the intensity and d is the distance to the membrane in micrometers.

We found an inverse relationship between the intensity slopes, m, and the PEA-determined thickness of the PCC, t ($S = m \cdot t$): Smaller PCCs exhibit steeper slopes (Fig. 2c).

As increased HA concentration lead to higher viscosities, the observed radial distribution of hyaluronan should be reflected by the mechanical properties of the PCC. Therefore, we performed corresponding microrheological measurements.

<u>Microrheology within the PCC.</u> The micromechanical properties of the PCC were analyzed via the Brownian motion of 0.45 μm polystyrene particles. In order to prevent unspecific

21

adsorption to the tracers we used carboxylated particles that should not interact with HA due to electrostatic repulsion. This could be confirmed in control microrheology (MR) experiments with HA solutions, which correspond well to literature values.[23] Additionally, observation of tracer particles over longer periods of time showed, that particles are able to diffuse into the PCC and out of it again.

In order to map the micromechanical properties of the PCC, we determined the position of each tracer particle in respect to the membrane and the edge of the PCC in terms of its relative position and analyzed its motion individually. The typical area covered by a tracer within the PCC calculated as the radius of gyration was 0.45 µm (st. dev. 0.24 µm), where this value also depends on the relative position of the particle within the PCC (Fig. 4a).

One difficulty in measuring aquaeous solutions, lies in the short time window in which particles stay within the field of view. They do not only move fast in x/y, but also in z leading to relativly short trajectories. Image acquisition with a high-speed camera enabled to set the minimal length of each track to 650 frames, thus ensuring sufficient / adequate statistics. To prove the quality of our analysis we compared the average microrhelogy values with the weighted slope α_w and MSD. The weighted tracks were averaged over a relative position of 0.1 and correspond very well to the values averaged over the same position (fig. 4 b,c). This indicates, that the chosen minimal particle track length is sufficient.

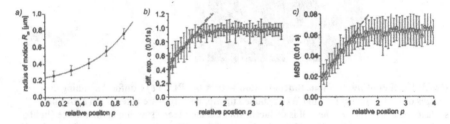

Fig. 4: a) Beads move less in closer proximity to the cell according to the increasing viscoelasticity, the radius of motion decreases exponentially towards the cell. The orange line depicts the exponential fit: $y = 0.15 + 009e^{0.48}$. The weighted data corresponds well to the non-weighted: The diffusive exponent shows the same curve progression for weighted (orange, squares) and averaged (blue circles) diffusive exponents (b) and MSDs (c), where the error bars depict the standard deviation of the averages. The linear fit performed between $0 < p < 1$, also overlap nicely: b) $y_w = 0.43 + 0.51x$ compared to $y_a = 0.47 + 0.47x$; c) $y_w = 0.014 + 0.039$ and $y_a = 0.018 + 0.036x$.

Inside the cell coat, variation of the viscoelasticity is indicated by slower motion of tracer particles and their correspondingly smaller MSDs as they approach the cell surface (Fig. 4c). These position-resolved MSD values clearly indicate a decrease in viscosity towards the edge of the PCC. Additionally, the diffusive exponent α, corresponding to the log-log slope of the MSD, also deviates from purely viscous diffusion inside the PCC Fig 4b).

DISCUSSION

We successfully employed microrheology in a position sensitive manner to obtain a profile of the pericellular coat. The observed viscoelasticity as well as the fluorescent intensity of GFPn labeled hyaluronan decrease from the cell membrane towards the rim of the PCC.

In the unique case of hyaluronan, this polysaccharide is synthesized directly on the surface.[24,25] If all HA would stay attached to these synthases and their concentration was high enough, the HA would stretch out to form a brush. As the HA loses entropic energy as it is stretched out, a force is required to pack the HA chains densly. This force could partially be provided by the synthase's extrusion forces. Such an end-grafted HA-brush might especially be formed in the hedghog cells induced by overexpression of one HA synthase[26] or in model-systems based on endgrafted HA.[27] Additionally, hyaladherins might play a vital role in such a brush-formation, as they stretch the HA.[7] The curve profile observed by our MR measurements as well as the HA stainings does however not fit to this end-grafted polymer brush, as it does not follow the predicted parabolic curve profile for such a system.

The HA within the PCC is also attached to the membrane by different receptors binding within the HA chain.[3] The attachment anywhere within the hyaluronan chain to membrane receptors could lead to a statistical distribution of effective chain lengths (Fig. 5) resulting in a more linear decreasing conc. profile even if the contour length of HA had a pretty long minimal length. In terms of polymer physics such a set-up would be best compared to the adhesive carpet model, which predicts a similar curve progression as observed in our experiments.

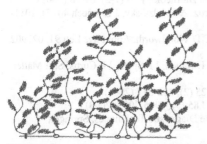

Fig. 5: Proposed mesoscopic structure of the PCC: All hyaluronan within the PCC is attached to the cell membrane either via a synthase (blue) or through membrane receptors (green) and is stretched out by hyaladherins like aggrecan in the case of chondrocytes.

In conclusion the PCC is most likely built up by hyaluronan attached to synthases and receptors within the cell membrane stretched out by hyaladherins like aggrecan. In contrast to multi-layered systems, this architecture could enable the cell to have a well defined 'handle' on its PCC through the actin linked HA-receptors within the cell membrane.

ACKNOWLEDGMENTS

We greatly appreciate the gift of RCJ-P cells from Prof. B. Geiger (Weizman Inst., Rehovot, Israel) and GFPn expressing cells from Uwe Rauch (Lund University, Sweden). We would like to give our thanks to Viktor Breedveld for inspiring discussions and the Nikon Center for technical assistance. JEC was partially funded by an Humboldt fellowship and CHJB by a Leopoldina scholarship.

REFERENCES

[1] S P Evanko, M Tammi, R H Tammi, T N Wight, Adv Drug Deliv Rev 59, 1351-1365 (2007).
[2] B P Toole, Nat Rev Cancer 4 (7), 528-539 (2004) .
[3] M Tammi, A J Day, E A Turley, J Biol Chem 277 (7), 4581-4584 (2002).
[4] N Itano et al., J Biol Chem 274 (35), 25085-25092 (1999).
[5] A J Day, G D Prestwich, J Biol Chem 277 (7), 4585-4588 (2002) .
[6] C B Knudson, J Biol Chem 120 (3), 825-834 (1993).
[7] M Hellmann, M Weiss, D W Heermann, Phys Rev E 76, 021802 (2007) .
[8] M Cohen, D Joester, I Sabanay, L Addadi, B Geiger, Soft Matter 3, 327–332 (2007) .
[9] B P Toole, Sem Cell Dev Bio 12 (2), 79-87 (2001).
[10] P M Freeman, R N Natarajan, J H Kimura, T P Andriacchi, J Orthop Res 12 (3), 311-320 (1994).
[11] M Knight, S Ghori, D Lee, D L Bader, Med Eng Phys 20, 684-688 (1998).
[12] W R Trickey, F P T Baaijens, T A Laursen, L G Alexopoulos, F Guilak, J Biomech 39 (1), 78-87 (2006).
[13] L G Alexopoulos, G M Williams, M L Upton, L A Setton, F Guilak, J Biomech 38 (3), 509-517 (2005) .
[14] D L Bader, T Ohashi, M Knight, D A Lee, M Sato, Biorheology 39 (1-2), 69-78 (2002).
[15] L Ng, H-H Hung, A Sprunt, S Chubinskaya, C Ortiz, A Grodzinsky, J Biomech 40 (5), 1011-1023 (2007).
[16] I Sokolov, S Iyer, V Subba-Rao, R M Gaikwad, C D Woodworth, Appl Phys Lett 91, 023902 (2007) .
[17] Boehm, T A Mundinger, C H J Boehm, V Hagel, U Rauch, J P Spatz, J E Curtis, Soft Matter 5 (21), 4331-4337 (2009) .
[18] M Cohen, Z Kam, L Addadi, B Geiger, EMBO J 25 (2), 302-311 (2006).
[19] M Cohen, E Klein, B Geiger, L Addadi, Biophys J 85 (3), 1996-2005 (2003).
[20] M. Gardel, M. Valentine and D. Weitz, Microscale Diagnostic Techniques, Springer, Heidelberg, 2005, Microrheology chapter.
[21] T A Waigh, Rep Prog Phys 68, 685-742 (2005).
[22] H Zhang, S L Baader, M Sixt, J Kappler, U Rauch, J Histochem Cytochem 52 (7), 915-922 (2004).
[23] H Boehm, PhD Thesis, University Heidelberg (2008).
[24] Weigel und DeAngelis, J Biol Chem 282 (51), 36777-36781 (2007) .
[25] DeAngelis, CMLS 56 (7), 670-682 (1999).
[26] A Kultti, K Rilla, R Tiihonen, A P Spicer, R H Tammi, M Tammi, J Biol Chem 281 (23), 15821-15828 (2006).
[27] R Richter, K Hock, J Burkhartsmeyer, H Boehm, P Bingen, G Wang, N F Steinmetz, D J Evans, J P Spatz, J Am Chem Soc 129, 5306-5307 (2007).

Mater. Res. Soc. Symp. Proc. Vol. 1274 © 2010 Materials Research Society   1274-QQ02-06

The contribution of the N- and C- terminal domains to the stretching properties of intermediate filaments

Laurent Kreplak
Department of Physics and Atmospheric Science, Dalhousie University, Halifax, NS B3H 3J5, Canada

ABSTRACT

The animal cell cytoskeleton consists of three interconnected filament systems: actin containing microfilaments (MFs), microtubules (MTs), and intermediate filaments (IFs). Among these three filaments systems, IFs are the only one that show high extensibility at both the single filament and network levels. In this work, I am presenting a simple model of IFs extensibility based on the current structural knowledge of the filaments. The only extra information added to this model compared to previous ones is the fact that the unfolded N- and C-termini of IF proteins are sandwiched between adjacent coiled-coil rod domains within the filaments. Since we know the contour length and typical persistence length of these unfolded termini, it is possible to predict the persistence length of a single filament, its maximal extensibility and the onset of coiled-coil unfolding. The predictions of the model are in good agreement with experiments on single desmin IFs stretched on a surface by AFM and on vimentin and desmin networks probed by rheology.

INTRODUCTION

Over the last twenty years the intermediate filaments (IFs) cytoskeleton has emerged as a key player in imparting mechanical stability to cells and tissues. The key feature of IF networks compared to actin and microtubule ones are their extensibility and ability to strain-harden [1]. However the molecular origin of these unique mechanical properties remains unclear [2]. One issue has been the lack of structural data on IF proteins and on the way they assemble into filaments. Most of the available structural information is confined to the well conserved central alpha-helical coiled-coil rod domain of IF proteins. The rod domain is also flanked by N- and C-terminal domains that should behave as unfolded polypeptide based on sequence analysis. These domains are highly variable among IFs but their presence is critical for proper filament assembly [3]. Using electron paramagnetic resonance spectroscopy, Aziz et al. demonstrated that the N-terminal domain of vimentin binds to its rod domain in a phosphorylation dependent manner [1]. Based on this new finding, I develop a structural model of IF mechanics that treats the N- and C-termini as entropic spring following the idea of Buehler and Ackbarow [4] . So far there are two main models of IF mechanics based on hard alpha-keratin fibers data [5] and on hagfish slime threads data [6]. The first one only takes into account the mechanical response of the rod domains, whereas the second one treats the terminal domains as entropic springs in series with the rod domains. The model presented here builds on the idea of Fudge et al. [6] taking into account the most recent structural data [1].

THEORY

The model is built for a murine desmin IF containing $n = 46$ chains per cross-section with a radius R of 6.3 nm [7]. Each desmin coiled-coil rod domain has a length b of 45 nm [3]. So a filament of length L_0 has N structural repeats with $N \approx \dfrac{L_0}{b}$. Each repeat contains 46 N-terminal domains and 46 C-terminal domains that, I assume, are sandwiched between adjacent rod domains (Figure 1). Since there are many ways for unfolded domains to be confined between the rod domains, I consider that only the last amino-acid of each N- or C- terminal domain is attached to an adjacent coiled-coil. Because the coiled-coil rod domains are half-staggered within the filament, I propose that each structural units contains two sets of springs mounted in series.

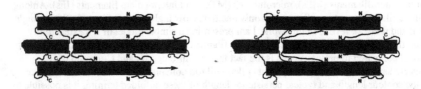

Figure 1. Structural model of a single IF. The rod domains are represented by solid rectangles and the N- and C- terminal domains are represented by thin lines. Sliding the rod domains past each-other is equivalent to stretching the unfolded terminal domains.

As a first approximation, I consider that each set is composed in average of 23 N-terminal domains and 23 C-terminal domains mounted in parallel. Stretching these springs is equivalent to sliding the rod domains past each-others. My next assumption is that the N- and C- terminal domains are more compliant than the coiled-coil rod domains and will have to extend first. In the case of murine desmin, the N-terminal domain has 108 amino-acids and it can be modeled as a worm-like chain with a persistence length λ of 0.43 nm [8] and a contour length L_{cN} of 41 nm. Similarly, the C-terminal domains, of 53 amino-acids each, should behave as worm-like chains with the same persistence length and a contour length L_{cC} of 20 nm. For double stranded coiled-coils, a typical value of the persistence length is 25 nm [9] so the rod domains are one to two orders of magnitude stiffer than an unfolded polypeptide. For extensions smaller than 90% of L_c, the N-terminal domain can be approximated by a spring of stiffness $k_N = \dfrac{k_B T}{\lambda L_{cN}}$ equal to 0.22 mN/m and the C-terminal can be approximated by a spring of stiffness $k_C = 0.46$ mN/m [10], $k_B T = 4$ pN.nm at room temperature. I can then replace each structural repeat by an entropic spring of stiffness $k_T = 7.8$ mN/m. The stiffness of a full filament of radius R made of N structural repeat is $k_{fil} = \dfrac{k_T}{N}$ and the equivalent modulus is $G_{fil} = \dfrac{k_{fil} L_0}{\pi R^2} = \dfrac{k_T b}{\pi R^2}$. In the case of a murine desmin filament $G_{fil} = 2.8$ MPa. I use the notation G because this is indeed a shear modulus and not a young's modulus. Stretching the entropic springs in our model is equivalent to a shearing process (see Figure 1). Still, since the N- and C- terminal domains are much more compliant than the coiled-coil domains, they should be responsible for the thermal fluctuations of the filament. In that case, the modulus G_{fil} and the persistence length λ_{fil} are linked through the following relationship $G_{fil}I = k_BT\lambda_{fil}$ with $I = \dfrac{\pi R^4}{4}$, assuming the filament is an homogeneous cylinder [11]. For a murine desmin filament, λ_{fil} is then equal to 870 nm.

The next step is to estimate the maximal extensibility of a filament. This is defined by the extensibility of the two types of elements in series, the terminal domains and the coiled-coil domains. From the atomic structure of a coiled-coil, the maximal extensibility is 2.5 folds the original length [12]. The terminal domains can't be stretched beyond their contour length. If we consider a single structural repeat along the filament, the largest length achievable is $L_{max} = L_{cN} + L_{cC} + 2.5b$, equal to 173 nm. This corresponds to a maximal extensibility of 3.8 folds the original length of the filament. This value is an overestimate, all the domains have a non linear mechanical response above 90% of their contour length which means that L_{max} should be limited to at most 156 nm (Figure 2). This corresponds to a maximal extensibility of 3.5 folds the original length of the filament.

The last mechanical information that can be extracted from this model is the onset of coiled-coil domains unfolding. This can only happen when a stretched terminal domain is loaded to a force equal or superior to the unfolding force of a coiled-coil. Based on AFM measurements of single myosin coiled-coils the unfolding force is larger than 25 pN [9,13]. Using the worm-like chain model again [10], a terminal domain must be stretched at least to 60% of its contour length to reach a 25 pN load (Figure 2). This leads to an estimated onset for coiled-coil unfolding inside the filament of 1.8 folds the original length. This value is most likely an overestimate as I assumed the termini to be attached to an adjacent coiled-coil by only their last amino-acid. If the termini are attached at multiple points then the onset may be as low as 15% extension [4].

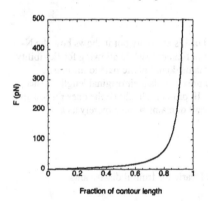

Figure 2. Computed force versus extension relation of a N- or C-terminal domain using the worm-like chain model of entropic elasticity [10].

DISCUSSION

The N- and C-terminal domains of IF proteins are already known to control the assembly and polymorphism of IFs [3]. However the mechanical role played by these domains is not well understood. The model presented above and derived from the one originally proposed by Fudge et al. [6], provides a good estimate for the persistence length of a desmin IF that has been measured by a high frequency rheology technique to 895 ± 50 nm [14]. It also provides a good estimate of the maximal extensibility of a single desmin IF as measured by AFM nanomanipulation [15]. Interestingly, a similar value of the extensibility was obtained recently from a steered molecular dynamics (MD) simulation of a vimentin tetramer under tension [16]. Except that in that case the N and C termini did not elongate. This may be due to the large pulling speed, in the order of 1 m/s, used to extend the tetramer.

The same model can be applied to a single human vimentin filament assuming a radius R of 5 nm, 32 chains per cross-section [17], a 60 residues long C-terminal domain and a 80 residues long N-terminal domain. Following the same reasoning as above gives $G_{fil} = 3.25$ MPa and $\lambda_{fil} = 400$ nm. This value of the persistence length is in good agreement with the ones determined experimentally by two different groups using rheology techniques [14,18]. Notice that G_{fil} and R do not change much between a desmin and a vimentin filament but the persistence length is divided by two. This points to the fact that λ_{fil} depends mainly on the number of chains per cross-section n. By realizing that $R \propto \sqrt{n}$, we get $\lambda_{fil} \propto n^2$. This simple scaling law fits well the available experimental data. Hence it is possible that cells control the stiffness of their IF network by regulating the number of chains per cross-section.

CONCLUSIONS

This is by no means a definite structural model of IFs elasticity but it shows how the N- and C- terminal domains may impart cohesiveness to the filaments while allowing for flexibility and extensibility. It also shows that the coiled-coil domains do not participate to the elastic response of the filaments until they are stretched by almost 2 folds their original length. Hence, in cells and tissues the N- and C- terminal domains of IF proteins should be the ones responsible for the observed elastic properties whereas the coiled-coil domains behave merely as a framework.

ACKNOWLEDGMENTS

I would like to thank Douglas Fudge and John Hearle for fruitful discussions.

REFERENCES

1. Aziz, A., Hess, J.F., Budamagunta, M.S., Voss, J.C., Fitzgerald, P.G., 2009. SDSL EPR determination of vimentin head domain structure. *J Biol Chem.* in press.
2. Kreplak, L., Fudge, D., 2007. Biomechanical properties of intermediate filaments: from tissues to single filaments and back. *Bioessays.* **29**, 26-35.
3. Kreplak, L., Aebi, U., Herrmann, H., 2004. Molecular mechanisms underlying the assembly of intermediate filaments. *Exp Cell Res.* **301**, 77-83.
4. Buehler, M. J., Acbarow, T., 2007. Fracture mechanics of protein materials. *Mat Today.* **10**,

46-58.

5. Hearle, J.W., 2000. A critical review of the structural mechanics of wool and hair fibres. *Int J Biol Macromol*. **27**, 123-138.

6. Fudge, D.S., Gardner, K.H., Forsyth, V.T., Riekel, C., Gosline, J.M., 2003. The mechanical properties of hydrated intermediate filaments: insights from hagfish slime threads. *Biophys J*. **85**, 2015-2027.

7. Bar, H., Mucke, N., Ringler, P., Muller, S.A., Kreplak, L., Katus, H.A., Aebi, U., Herrmann, H., 2006. Impact of disease mutations on the desmin filament assembly process. *J Mol Biol*. **360**, 1031-1042.

8. Lim, R.Y., Koser, J., Huang, N.P., Schwarz-Herion, K., Aebi, U., 2007. Nanomechanical interactions of phenylalanine-glycine nucleoporins studied by single molecule force-volume spectroscopy. *J Struct Biol*. **159**, 277-289.

9. Schwaiger, I., Sattler, C., Hostetter, D.R., Rief, M., 2002. The myosin coiled-coil is a truly elastic protein structure. *Nat Mater*. **1**, 232-235.

10. Bustamante, C., Marko, J.F., Siggia, E.D., Smith, S., 1994. Entropic elasticity of lambda-phage DNA. *Science*. **265**, 1599-1600.

11. Mucke, N., Kreplak, L., Kirmse, R., Wedig, T., Herrmann, H., Aebi, U., Langowski, J., 2004. Assessing the flexibility of intermediate filaments by atomic force microscopy. *J Mol Biol*. **335**, 1241-1250.

12. Staple, D.B., Loparic, M., Kreuzer, H.J., Kreplak, L., 2009. Stretching, unfolding, and deforming protein filaments adsorbed at solid-liquid interfaces using the tip of an atomic-force microscope. *Phys Rev Lett*. **102**, 128302.

13. Root, D.D., Yadavalli, V.K., Forbes, J.G., Wang, K., 2006. Coiled-coil nanomechanics and uncoiling and unfolding of the superhelix and alpha-helices of myosin. *Biophys J*. **90**, 2852-2866.

14. Schopferer, M., Bar, H., Hochstein, B., Sharma, S., Mucke, N., Herrmann, H., Willenbacher, N., 2009. Desmin and vimentin intermediate filament networks: their viscoelastic properties investigated by mechanical rheometry. *J Mol Biol*. **388**, 133-143.

15. Kreplak, L., Herrmann, H., Aebi, U., 2008. Tensile properties of single desmin intermediate filaments. *Biophys J*. **94**, 2790-2799.

16. Qin, Z., Kreplak, L., Buehler, M. J., 2009. Hierarchical structure controls nanomechanical properties of vimentin intermediate filaments. *PloS ONE*. **4**, e7294.

17. Herrmann, H., Haner, M., Brettel, M., Muller, S.A., Goldie, K.N., Fedtke, B., Lustig, A., Franke, W.W., Aebi, U., 1996. Structure and assembly properties of the intermediate filament protein vimentin: the role of its head, rod and tail domains. *J Mol Biol*. **264**, 933-953.

18. Lin, Y.C., Yao, N.Y., Broedersz, C.P., Herrmann, H., Mackintosh, F.C., Weitz, D.A., 2010. Origins of elasticity in intermediate filament networks. *Phys Rev Lett*. **104**, 058101.

Ectopic Materials in the Context of Prion and Amyloid Diseases

Mater. Res. Soc. Symp. Proc. Vol. 1274 © 2010 Materials Research Society 1274-QQ04-05

Quantitative approaches for characterising fibrillar protein nanostructures

Duncan A. White[a], Christopher M. Dobson[a], Mark E Welland[b] and Tuomas P. J. Knowles[a]

[a]Department of Chemistry, University of Cambridge, Lensfield Road, Cambridge, CB2 1EW, United Kingdom.

[b]Nanoscience Centre, University of Cambridge, J. J. Thomson Avenue, Cambridge, CB3 0FF, United Kingdom.

Abstract

Polypeptide sequences have an inherent tendency to self-assemble into filamentous nanostructures commonly known as amyloid fibrils. Such self-assembly is used in nature to generate a variety of functional materials ranging from protective coatings in bacteria to catalytic scaffolds in mammals. The aberrant self-assembly of misfolded peptides and proteins is also, however, implicated in a range of disease states including neurodegenerative conditions such as Alzheimer's and Parkinson's diseases. It is increasingly evident that the intrinsic material properties of these structures are crucial for understanding the thermodynamics and kinetics of the pathological deposition of proteins, particularly as the mechanical fragmentation of aggregates enhances the rate of protein deposition by exposing new fibril ends which can promote further growth. We discuss here recent advances in physical techniques that are able to characterise the hierarchical self-assembly of misfolded protein molecules and define their properties.

Introduction

Amyloid fibrils are high aspect ratio protein nanostructures that are formed through the self-assembly of polypeptide chains. They are better known as a result of their association with debilitating disorders such as Alzheimer's disease and type II diabetes where normally soluble peptide and protein molecules misfold and aggregate into amyloid structures [1, 2]. These highly ordered fibrils are characterised by a 'cross-β' core structure formed from β-strands that are predisposed to form fibrillar aggregates and often consist of several protofilaments twisted around one another, resulting in extremely robust mechanical properties [3–5]. Although most natural amyloid structures have pathological associations, in a number of cases they may be found to have biologically functional roles [6, 7]. As a result of this situation, a detailed and quantitative characterisation of the material properties and growth kinetics of amyloid fibrils is essential to the development of novel and responsive multidimensional templates and biomaterials for future technological applications [8–13]. Such an approach aimed at bridging the nano and macro scales necessitates discussion of appropriate quantitative methods to probe the material properties of amyloid fibrils and the kinetics of their assembly.

In addition to material considerations, it is critically important to obtain quantitative measurements of fibril growth and breakage as these are key factors related to the proliferation of amyloid structure in many debilitating diseases [14–17]. This article therefore focuses on these two aspects of amyloid systems and demonstrates how new approaches can result in dramatic improvements in our ability to obtain quantitative data.

Material Properties

Despite the fact that amyloid fibrils can be formed from chemically and structurally diverse polypeptide sequences, they typically exhibit common and universal structural elements on the nanoscale such as the existence of protofilaments that may twist around one another (Figure 1a and b), and a characteristic X-ray fibre diffraction pattern [18, 19]. The latter indicates the existence of β-strands lying perpendicular to the fibril axis with hydrogen bonds aligned parallel to the fibril axis. This elegant structure provides outstanding mechanical strength attributed to a highly ordered, coherent, and intricate arrangement of intermolecular bonding that dominates over intramolecular interactions as the soluble protein misfolds and aggregates [4, 12, 13].

The atomic force microscope (AFM) has proved to be an invaluable tool for the mechanical characterisation of individual amyloid fibrils and will be the focus of our discussions here. Pioneering biomechanical AFM studies have demonstrated the ability to unfold native states of proteins and to provide important information about the forces that determine these states [20, 21]. In the case of individual amyloid fibrils, it has been shown that by measuring directly the mechanical properties of these structures, including stiffness and buckling, it is possible to determine material characteristics through statistical analysis of filament shapes [3–5], and to elucidate fibril growth kinetics with high temporal resolution [22].

The mechanical properties of materials are commonly measured by applying a load in a three-point bending geometry, such as when a beam is suspended from both sides and a known force applied. Nanotechnology has enabled similar tests to be performed upon amyloid fibrils [3] where a cantilever is used to apply a load on to a fibril suspended over grooves etched into a silicon substrate. The deflection of the cantilever is then used as a measure of the applied force, and as the degree of bending can be well defined; the Young's modulus may then be calculated from the resulting force-distance curve. For two-filament insulin fibrils this value was reported to be 3.3 ± 0.4 GPa which is on the same order as spider silk [23].

An additional AFM approach to determine the material properties of amyloid fibrils involves the analysis of AFM images of amyloid fibrils adsorbed to surfaces. This method involves analysis of the decay of tangent correlations as a result of shape fluctuations along a fibril axis and has been used to determine the persistence length and bending (or flexural) rigidity of individual amyloid fibrils from a series of different proteins [4, 24]. Remarkably, it was found that the bending rigidities for the fibrils of the different proteins varied over four orders of magnitude (from $C_B = 1.4 \times 10^{-28}$ Nm^2 to $C_B = 1.3 \times 10^{24}$ Nm^2). The experimentally determined bending rigidity for Aβ(1-42), the peptide implicated in Alzheimer's disease, was corroborated by atomistic

34

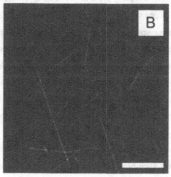

Figure 1: Amyloid fibrils. (a) 3D reconstructions of 2 and 3 protofilament insulin fibrils from cryo-electron microscopy data [39]. (b) Tapping mode atomic force microscopy image of transthyretin (105-115) fibrils on a mica surface, the scale bar represents 1 μm [4].

simulations of the nanomechanics of the closely related Aβ(1-40) peptide [25].

The inherent timescale and spatial resolution of AFM means that the technique is not limited to structural and mechanical observations of pre-grown fibrils but is also capable of accurate kinetic measurements of amyloid fibril growth. The in situ growth kinetics of an amylin protofibril on a mica surface was monitored by AFM in real time and observed to proceed with an elongation rate of 1.1 ($\pm$0.5) nm min^{-1} [26]. Similar studies on the elongation of a protofibril from the peptide, Aβ(1-40), yielded elongation rates varying from 1 to 10 nm min^{-1} [27]. A further study observed polarized and stepwise assembly dynamics for a fragment of the Aβ(1-42) peptide, Aβ(25-35), and revealed an exponential distribution of "pause steps" (time constant, 32.3 s $\pm$ 0.8 s) interspersed with phases of rapid growth with elongation rates of 262.2 $\pm$ 0.54 nm min^{-1} [22].

The kinetics of fibril self-assembly

Assays for monitoring amyloid fibril growth kinetics in vitro traditionally involve the use of amyloid specific dyes such as Congo red and thioflavin T. The binding stoichiometries of such dyes are challenging to quantify and this difficulty is accentuated by the relatively low specificity of such dyes for amyloid structures [28, 29]. These observations have motivated the development of label-free methods for the investigation of fibril elongation, the most successful of which have been surface plasmon resonance (SPR) and the quartz crystal microbalance (QCM) [30–32]. We shall focus on the latter in these discussions as the method has perhaps received less attention than SPR yet offers a unique perspective to challenge or prove existing paradigms of the mechanism of amyloid function.

Measurements of amyloid fibril growth made with QCM rely upon the high mass sensitivity at a solid-liquid interface of an oscillating quartz crystal in contact with a liquid [33]. QCM, therefore, is intrinsically limited to reactions taking place at the solid-liquid interface and as such must involve functionalisation of a surface with a precursor species; in the current case either monomeric protein or seed fibrils. The surface-bound precursor species are then exposed to soluble protein under denaturing conditions and the growth of amyloid fibrils is monitored as a decrease in the resonant frequency of a quartz crystal. Modern QCM systems are coupled to fluid handling devices allowing high-throughput analysis with measurement times for a typical elongation based assay on the order of minutes.

Templated growth from surface-bound seed fibrils has enabled measurement of the elongation rates of amyloid fibrils from a diverse set of protein molecules, including insulin, lysozyme, the $A\beta$ peptide and α-synuclein, under a variety of conditions [30, 34–36]. In particular, the method has allowed the elongation kinetics of amyloid fibril growth to be studied without contributions from stochastic processes such as nucleation and fibril breakage, resulting in much higher accuracy and reproducibility over traditional methods [30, 36]. These studies have recently been extended, demonstrating the versatility of the technique to make accurate kinetic measurements in complex crowded environments where it is very difficult to achieve robust observations using traditional dye-based assays. Specifically, it was found that in the presence of a variety of concentrations of "crowding molecules" subtle modulation of aggregation kinetics (within a factor of two) was easily resolved and found to be in excellent agreement with predictions from existing theoretical descriptions of fibril elongation in crowded environments [36, 37]. Furthermore, QCM measurements have provided sufficient range and resolution to allow measurement of changes in fibril elongation rates, that span four orders of magnitude, arising from single point mutations [35].

The nucleation stage of the aggregation process is a vital precursor of fibril formation, growth and fragmentation. A clear description of this process is therefore an extremely important aspect of protein aggregation and yet high-quality experimental data remain elusive. Some studies have, however, extended the application of the QCM to study nucleation events from multilayers of surface adsorbed monomeric protein molecules [38]. Exposure of a suitably functionalised surface to high concentrations of soluble protein can result in multilayer formation under denaturing conditions from which multiple nucleation events and subsequent fibril growth could be observed. Such pioneering studies are pushing the boundaries forward yet further of the QCM method and have potential to provide unique insights into the process of protein aggregation. Additionally, the equipment needed for these experiments is commercially available and the short times needed for an assay compared to typical dye-based methods provides an excellent opportunity for high throughput measurements and screening.

Conclusion

The material properties and self-assembly kinetics of amyloid fibrils are fundamental to a complete understanding of the aggregation process implicated in many degenerative diseases and also for a firm foundation upon which to base future research in novel biomaterials. It is only recently that accurate methods have been developed to probe quantitatively the key parameters that govern amyloid fibril growth but they have already had significant impacts on the field and present many promising avenues for future research.

References

[1] Dobson, C. M. (1999) Protein misfolding, evolution and disease. Trends Biochem. Sci. 24(9), 329–332.

[2] Dobson, C. M. (2003) Protein folding and misfolding. Nature 426(6968), 884–890.

[3] Smith, J. F., Knowles, T. P. J., Dobson, C. M., Macphee, C. E. and Welland, M. E. (2006) Characterization of the nanoscale properties of individual amyloid fibrils. Proc Natl Acad Sci U S A 103(43), 15806–15811.

[4] Knowles, T. P., Fitzpatrick, A. W., Meehan, S., Mott, H. R., Vendruscolo, M., Dobson, C. M. and Welland, M. E. (2007) Role of intermolecular forces in defining material properties of protein nanofibrils. Science 318(5858), 1900–1903.

[5] Dong, M. D., Hovgaard, M. B., Mamdouh, W., Xu, S. L., Otzen, D. E. and Besenbacher, F. (2008) Afm-based force spectroscopy measurements of mature amyloid fibrils of the peptide glucagon Nanotechnology 19(38), 384013.

[6] Fowler, D. M., Koulov, A. V., Alory-Jost, C., Marks, M. S., Balch, W. E. and Kelly, J. W. (2006) Functional amyloid formation within mammalian tissue. PLoS Biol. 4(1), e6.

[7] Fowler, D. M., Koulov, A. V., Balch, W. E. and Kelly, J. W. (2007) Functional amyloid - from bacteria to humans. Trends Biochem. Sci. 32(5), 217–224.

[8] Scheibel, T., Parthasarathy, R., Sawicki, G., Lin, X.-M., Jaeger, H. and Lindquist, S. L. (2003) Conducting nanowires built by controlled self-assembly of amyloid fibers and selective metal deposition. Proc Natl Acad Sci U S A 100(8), 4527–4532.

[9] Reches, M. and Gazit, E. (2003) Casting metal nanowires within discrete self-assembled peptide nanotubes. Science 300(5619), 625–627.

[10] Pilkington, S. M., Roberts, S. J., Meade, S. J. and Gerrard, J. A. (2010) Amyloid fibrils as a nanoscaffold for enzyme immobilization. Biotechnol Prog 26(1), 93–100.

[11] Knowles, T. P. J., Oppenheim, T. W., Buell, A. K., Chirgadze, D. Y. and Welland, M. E. (2010) Nanostructured films from hierarchical self-assembly of amyloidogenic proteins. Nat Nanotechnol 5(3), 204–207.

[12] Buehler, M. J. and Yung, Y. C. (2010) How protein materials balance strength, robustness, and adaptability Hfsp Journal 4(1), 26–40.

[13] Buehler, M. J. (2010) Nanomaterials: Strength in numbers Nat Nano 5(3), 172–174.

[14] Knowles, T. P. J., Waudby, C. A., Devlin, G. L., Cohen, S. I. A., Aguzzi, A., Vendruscolo, M., Terentjev, E. M., Welland, M. E. and Dobson, C. M. (2009) An analytical solution to the kinetics of breakable filament assembly. Science 326(5959), 1533–1537.

[15] Tanaka, M., Collins, S. R., Toyama, B. H. and Weissman, J. S. (2006) The physical basis of how prion conformations determine strain phenotypes. Nature 442(7102), 585–589.

[16] Collins, S. R., Douglass, A., Vale, R. D. and Weissman, J. S. (2004) Mechanism of prion propagation: amyloid growth occurs by monomer addition. PLoS Biol 2(10), e321.

[17] Nowak, M. A., Krakauer, D. C., Klug, A. and May, R. M. (1998) Prion infection dynamics Integrative Biology.

[18] Chiti, F. and Dobson, C. M. (2006) Protein misfolding, functional amyloid, and human disease. Annu. Rev. Biochem. 75, 333–366.

[19] Paparcone, R. and Buehler, M. J. (2009) Microscale structural model of alzheimer a beta(1–40) amyloid fibril Applied Physics Letters 94(24), 243904.

[20] Rief, M., Gautel, M., Oesterhelt, F., Fernandez, J. M. and Gaub, H. E. (1997) Reversible unfolding of individual titin immunoglobulin domains by afm. Science 276(5315), 1109–1112.

[21] Carrion-Vazquez, M., Oberhauser, A. F., Fowler, S. B., Marszalek, P. E., Broedel, S. E., Clarke, J. and Fernandez, J. M. (1999) Mechanical and chemical unfolding of a single protein: a comparison. Proc Natl Acad Sci U S A 96(7), 3694–3699.

[22] Kellermayer, M. S. Z., Karsai, A., Benke, M., Soós, K. and Penke, B. (2008) Stepwise dynamics of epitaxially growing single amyloid fibrils. Proc Natl Acad Sci U S A 105(1), 141–144.

[23] Vollrath, F. and Knight, D. P. (2001) Liquid crystalline spinning of spider silk Nature 410(6828), 541–548.

[24] Adamcik, J., Jung, J.-M., Flakowski, J., Rios, P. D. L., Dietler, G. and Mezzenga, R. (2010) Understanding amyloid aggregation by statistical analysis of atomic force microscopy images. Nat Nanotechnol.

[25] Paparcone, R., Keten, S. and Buehler, M. J. (2010) Atomistic simulation of nanomechanical properties of alzheimer's abeta(1-40) amyloid fibrils under compressive and tensile loading. J Biomech 43(6), 1196–1201.

[26] Goldsbury, C., Kistler, J., Aebi, U., Arvinte, T. and Cooper, G. J. (1999) Watching amyloid fibrils grow by time-lapse atomic force microscopy. J Mol Biol 285(1), 33–39.

[27] Blackley, H. K., Sanders, G. H., Davies, M. C., Roberts, C. J., Tendler, S. J. and Wilkinson, M. J. (2000) In-situ atomic force microscopy study of beta-amyloid fibrillization. J Mol Biol 298(5), 833–840.

[28] Khurana, R., Uversky, V. N., Nielsen, L. and Fink, A. L. (2001) Is Congo red an amyloid-specific dye? J. Biol. Chem. 276(25), 22715–22721.

[29] Krebs, M. R. H., Bromley, E. H. C. and Donald, A. M. (2005) The binding of thioflavin-T to amyloid fibrils: localisation and implications. J. Struct. Biol. 149(1), 30–37.

[30] Knowles, T. P. J., Shu, W., Devlin, G. L., Meehan, S., Auer, S., Dobson, C. M. and Welland, M. E. (2007) Kinetics and thermodynamics of amyloid formation from direct measurements of fluctuations in fibril mass. Proc. Natl. Acad. Sci. USA 104(24), 10016–10021.

[31] Okuno, H., Mori, K., Jitsukawa, T., Inoue, H. and Chiba, S. (2006) Convenient method for monitoring Abeta aggregation by quartz-crystal microbalance. Chem. Biol. Drug. Des. 68(5), 273–275.

[32] Myszka, D. G. (1997) Kinetic analysis of macromolecular interactions using surface plasmon resonance biosensors. Curr. Opin. Biotech. 8(1), 50–57.

[33] Kanazawa, K. K. and Gordon, J. G. (1985) Frequency of a quartz microbalance in contact with liquid. Anal. Chem. 57(8), 1770–1771.

[34] Kotarek, J. A., Johnson, K. C. and Moss, M. A. (2008) Quartz crystal microbalance analysis of growth kinetics for aggregation intermediates of the amyloid-beta protein. Anal. Biochem. 378(1), 15–24.

[35] Buell, A. K., Tartaglia, G. G., Birkett, N. R., Waudby, C. A., Vendruscolo, M., Salvatella, X., Welland, M. E., Dobson, C. M. and Knowles, T. P. J. (2009) Position-dependent electrostatic protection against protein aggregation. Chembiochem 10(8), 1309–1312.

[36] White, D. A., Buell, A. K., Knowles, T. P. J., Welland, M. E. and Dobson, C. M. (2010) Protein aggregation in crowded environments. J Am Chem Soc 132(14), 5170–5175.

[37] Minton, A. P. (1983) The effect of volume occupancy upon the thermodynamic activity of proteins: some biochemical consequences. Mol Cell Biochem 55(2), 119–140.

[38] Hovgaard, M. B., Dong, M., Otzen, D. E. and Besenbacher, F. (2007) Quartz crystal microbalance studies of multilayer glucagon fibrillation at the solid-liquid interface. Biophys. J. 93(6), 2162–2169.

[39] Jiménez, J. L., Nettleton, E. J., Bouchard, M., Robinson, C. V., Dobson, C. M. and Saibil, H. R. (2002) The protofilament structure of insulin amyloid fibrils. Proc Natl Acad Sci U S A 99(14), 9196–9201.

Poster Session

Mater. Res. Soc. Symp. Proc. Vol. 1274 © 2010 Materials Research Society   1274-QQ05-01

Controlling the physical properties of random network based shape memory polymer foams

Pooja Singhal[1], Thomas S. Wilson[2] and Duncan J. Maitland[1, 2]
[1] Texas A&M University, Biomedical Engineering Department. 337 Zachry Engineering Center, MS 3120, College Station Texas- 77843.
[2] Lawrence Livermore National Laboratory, 7000 East Avenue, Livermore, CA- 94551.

ABSTRACT

Shape memory polymers (SMPs) are increasingly being considered for use in minimally invasive medical devices. For safe deployment of implanted devices it is important to be able to precisely control the actuation temperature of the device. In this study we report the effect of varying monomer composition on the glass transitions/actuation temperatures (Tg) of novel low density shape memory foams. The foams were based on hexamethylenediisocyanate (HDI), triethanolamine (TEA) and tetrakis (2-hydroxyl propyl) ethylenediamine (HPED), and were produced via a combination of chemical and physical blowing process. The process for post-foaming cleaning was also varied. Foams were characterized by DSC, DMA, and for shape memory. No clear trends were observed for foam samples without cleaning, and this was attributed to process chemicals acting as plasticizers. In foams cleaned via washing and/or sonication, the Tg was observed to decrease for compositions that were higher in the TEA content. Also, no change in shape memory behavior was observed for varying compositions. This work demonstrates the ability to tailor actuation transition temperature while maintaining shape memory behavior for low density foams suitable for aneurysm treatment.

INTRODUCTION

Shape memory polymers are synthetic polymeric materials that can be deformed to a stable secondary shape from their original primary shape. They retain the secondary shape until an appropriate stimulus is provided. Their response to a variety of stimuli, i.e. light, heat, electricity etc. has been reported [1-3]. Amongst these, thermally responsive SMPs are the most widely studied polymers.

Thermal shape-memory behavior has been observed in a number of polymer systems comprised of covalently crosslinked amorphous or semi-crystalline networks as well as physically cross-linked glassy or semi-crystalline block copolymers [4]. Considering a potential deployment in human body, several safer methods of heating the SMPs have been investigated as possible triggering mechanisms for these materials, for e.g. electric, electromagnetic and radio frequency [5-7]. But still, thermal damage to tissue is a concern in medical procedures, and it is best to minimize it. Hence, owing to their promising future in biomedical applications [8], controlling the actuation temperature to be close to body temperature is an important requirement for safe deployment of biomedical devices based on these SMPs.

Here we report the effect of changing monomer functionality on the shape memory behavior of highly crosslinked amorphous polymers with a very regular network structure [9]. These materials are synthesized from hexamethylenediisocyanate (HDI), triethanolamine (TEA) and tetrakis (2-hydroxyl propyl) ethylenediamine (HPED). The network structure of polymer is formed due to tri- and tetra- functional TEA and HPED respectively. The shape memory behavior of these amorphous, highly crosslinked materials is related to their glass transition (Tg) [4]. In such systems, shape recovery is driven by entropy of the chains. Prior to actuation, the chains are in a non-equilibrium conformation with net orientation, which is stabilized by a lack of mobility due to the inability of chain segments to move relative to one another below the glass transition. By heating above Tg, the network chains become mobile and are entropically driven to their equilibrium conformation, resulting in macroscopic change in shape. The use of tri and tetra-functional monomers, with relatively low molecular weight difunctional monomer connecting them, creates a network with a high volume concentration of crosslink sites. These crosslinks act to prevent relaxation of material while under applied deformation, and likewise provide for a high recovery force in the material according to the well known relationship between modulus and crosslink density for crosslinked polymers in their rubbery state; namely, $E \sim n_c RT$, where E is modulus, n_c is volume concentration of crosslinks, R is the gas constant, and T is temperature [10]

Theoretically, tuning the glass transition for these polymers can be achieved either by using longer/more flexible individual molecules which will thereby decrease the crosslink density and hence the glass transition, or by changing the rotational stiffness and interchain enthalpy of those network chains. Here we have investigated the change in degree of chain mobility by varying the ratio of HPED and TEA in the polymer network. The hypothesis behind this work is that a higher TEA content, in comparison to HPED content, will show a higher mobility and lower actuation temperature, while maintaining a strong shape memory behavior.

EXPERIMENT

Synthesis
Three sets of prepolymers were prepared with varying ratios of TEA and HPED as shown in Figure 1. The prepolymers were formulated with an OH/NCO ratio 0.38 and allowed to cure for 48 hours at room temperature. Prefixes L, M and H in the names represent low, medium and high contents of HPED respectively.

Table 1: Composition of prepolymer

Sample name	HDI (wt%)	TEA (wt%)	HPED (wt%)
LHPED	77.7	7.1	15.3
MHPED	76.7	4.4	18.8
HHPED	76.1	2.6	21.4

Foaming was done by adding the balance of hydroxyl groups as TEA and de-ionized water to make an OH/NCO ratio 0.95 (105 isocyanate index formulation), along with surfactants and catalysts (Air Products Inc.) and physical blowing agent (Honeywell Corp.). Foams were prepared by the hand mixing procedure. Total mixing time, cream time, top of the cup time and end of rise time were measured for each foam sample prepared and the foams were allowed to cure at room temperature for 2 days before being analyzed.

Cleaning of foams
For cleaning, 8 mm diameter cylindrical samples of foam were cut out using a biopsy punch. In the first step, samples were soaked in DI water for 12 hours, followed by sonication in DI water for 30 minutes. In the second step, samples were sonicated in 0.1N Hydrochloric acid (HCl) for 2 hours and then were rinsed and sonicated in 80-20 volume % DI-Conrad solution for 1 hour. This was followed by multiple rinses in DI water until the water ran clear, and a final sonication in DI water for 1 hour.

Density
Density was measured for cylindrical samples of uncleaned foams. Weight was measured on a Mettler Toledo AB54-S scale to 0.1 mg resolution. Diameter and height of the cylindrical samples were measured using digital micrometer to calculate the volume. Density was then calculated from mass per unit volume.

Glass transition (Tg)
Tg was measured for the untreated samples, and for samples treated by the above two-step cleaning processes. A TA instruments Q-200 Differential Scanning Calorimeter (DSC) was first equilibrated at -40 °C, ramped up to 200 °C, back to -40°C and then again to 200 °C at 10 °C/minute. The half height of the transition of the second heating curve was used to determine the glass transition by DSC.

Shape memory behavior
The initial diameter of the cylindrical samples was measured. The crimper (Stent crimp equipment Machine Solutions SC150-42) was preheated to 100 °C for 1 hour, and was loaded with the sample. The sample was allowed to equilibrate for 20 minutes and then crimped with minimum (zero) diameter setting to maximize the crimping force. The temperature was then decreased and crimped foam was allowed to cool to room temperature by blowing cool (20 °C) air (blower, Beahm Designs) on it for one hour. The diameter of the crimped sample was measured using a digital micrometer. Thereafter the foam was re-expanded by blowing hot air (100 °C) for a few seconds. While actuation of the materials could be achieved close to their individual glass transition temperatures, a higher temperature was used here to compare the recovery of the different compositions at the same temperature. The expanded foam sample was allowed to cool down and recovered dimensions were recorded.

Modulus
The Storage and loss modulus, and tan delta curves of the materials were recorded using the TA Instruments RSA 3 Dynamic Mechanical Analyzer (DMA). Dynamic temperature ramp tests were run in compression mode at 0.08% strain and 1 Hz frequency. A sweep of 0-150 °C temperature range at a rate of 5 °C /minute was used.

RESULTS AND DISCUSSION

The variation in the formulation of prepolymers is expected to give a higher fraction of tetra-functional crosslinks due to higher HPED content in HHPED foams, as compared to those in LHPED foams.

The changes in the foaming parameters based on the formulation difference were not very significant. Increasing the amount of TEA slightly increased the total mixing time required, as well as the time of foam rise.

Table 2: Foaming parameters for the three foam formulations

	HHPED	MHPED	LHPED
Total mixing time (s)	72	84	94
Cream time (s)	25	23	30
Top of cup time (s)	-	65	98
End of rise time (s)	60	78	98

The density of the foams ranged from 0.04 to 0.06 g/cc. LHPED samples showed an average density of 0.058 g/cc, while MHPED and LHPED samples showed average densities of 0.041 and 0.055 g/cc respectively. No identifiable trend was observed in the density of the foam samples. Hence the density and cell structure were not determined to be a primary function of the constituent composition used here. The variability in the density values can be based both on the hand mixing procedure adopted for making these foams, and the possible difference in the rheological properties of the prepolymers.

The glass transition as measured from DSC showed much lower values than expected from earlier work [9]. This was explained from the possible presence of plasticizing agents i.e. residual surfactants, catalysts and blowing agents which can interfere with the packing of the polymer chains and allow for their mobility with lower amount of thermal energy. Also, the glass transitions in the foam samples did not show any trend, which is likely due to the lack of control over the amount of residual plasticizing species for different foams.

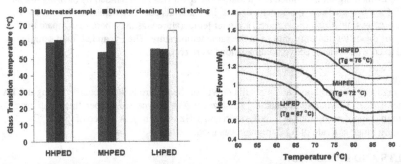

Figure 1: Variation in glass transition of the polymer samples due to changes in HPED vs. TEA content. (Left) Comparison of Tg between untreated samples and samples cleaned with DI water and 0.1 N Hydrochloric acid respectively. (Right) The Differential Scanning Calorimetry curves for heat flow for the respective samples.

Two levels of cleaning procedures were performed on the foam samples and changes in Tg were measured post cleaning. The mild acidic medium provided by 0.1 N HCl was seen to be a more effective method for this purpose compared to DI water. The expected trend of glass transition [9] was observed after performing the cleaning procedures. This indicates a significant fraction of mobile species present in the foams post-cure were removed by cleaning.

The percent shape recovery was high for all the HHPED, MHPED and LHPED foam samples without any noticeable difference. This can be explained from the fact that all these materials have a highly crosslinked structure. The subtle differences in the degree of crosslinking, based on monomer functionality, were shown to have an effect on other physical properties of the foams, like Tg. But the extent of shape recovery does not seem to get affected significantly by this parameter in the tested range.

Table 3: Study of shape memory behavior of the polymer foam samples.

Sample name	Initial diameter (mm)	Crimped diameter (mm)	Expanded diameter (mm)	Expansion (%)	Recovery (%)
HHPED	8.03	1.67	7.96	22.61	98
MHPED	8.11	1.51	8.04	28.2	98
LHPED	8.02	1.64	7.93	23.27	98

The storage modulus and tan delta curves of the uncleaned foam samples did not show any specific trend in the three samples. However, the cleaned samples again confirmed the difference in the chain mobility achieved via the difference in HPED vs. TEA concentrations and generated the expected trend of results [9].

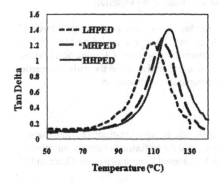 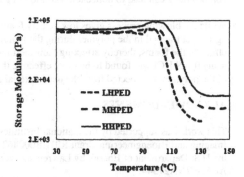

Figure 2: (Left) Tan Delta curve for the cleaned foam samples. The glass transition values are seen to decrease as we go from HHPED to LHPED foam samples. (Right) Similar trend is noticed in the change in storage modulus.

The values of glass transition as recorded from the DMA measurements gave difference of 5 and 3 °C respectively between the LHPED (110 °C), MHPED (115 °C) and HHPED (118 °C) samples. The difference in Tg values was the same in DSC and DMA results. However, there was a significant difference in the value of Tg itself in the two measurements; DMA gave much higher Tg values than DSC. This is explained from two factors- a) Firstly, the rate of temperature ramp (5 °C/minute) in DMA measurements. On reducing the temperature ramp rate to 2 °C/minute in the DMA tests, better agreement with DSC results was obtained. b) Secondly, the analyses of tan-delta peaks for Tg, as opposed to the peak of loss modulus curve. Tan-delta peak values are several degrees higher than the loss modulus peaks and represent the transition midpoint, which exceeds the "softening" point of the polymers [11]. Tg values from DSC are in general shown to be lower than those obtained from DMA [12].

Overall, the lower content of tetra- functional HPED was seen to allow for relaxation of chains at a lower temperature. Two important mechanisms could be playing a role here. The terminal methyl groups on the HPED can potentially reduce the mobility at the urethane linkage and cause the Tg of compositions with higher HPED content to get increased. Also since HPED contributes to four bonds with isocyanate per molecule, compared to three in case of TEA, a structure with higher volume concentration of crosslinks is expected to form with higher content of HHPED. This can lead to reduced chain mobility in these compositions, giving them a higher glass transition temperature compared to the material with lower HPED content.

CONCLUSIONS

Varying the ratio of TEA vs. HPED in highly branched shape memory polymers is an effective way of tuning the glass transition while maintaining high shape memory behavior of these SMPs. This is explained by both, the terminal methyl groups in HPED that can reduce mobility at urethane linkage, and a more closely packed network structure with higher content of tetra-functional HPED. A proof-of-concept study was performed to show that even slight variations in compositions can lead to noticeable variation in glass transition of the material.

While the blowing process gives low density foams, residual surfactants, catalysts and blowing agents get trapped in the polymer during the foaming process. These species have a plasticizing effect on the foams, thereby affecting their thermo-mechanical properties. Cleaning of the foams using 0.1 N HCl was found to be more effective than DI water, and restored the native properties of the polymer as expected from the neat material of comparable composition.

ACKNOWLEDGMENTS

This work was supported by the National Institutes of Health, National Institute of Biomedical Imaging and Bioengineering Grant R01EB000462 and partially performed under the auspices of the U.S. Department of Energy by Lawrence Livermore National Laboratory under Contract DE-AC52-07NA27344.

REFERENCES

1.   A. Lendlein, H. Y. Jiang, O. Junger and R. Langer, Nature **434**, 879–882 (2005).

2.   N. G. Sahoo, Y. C. Jung and J. W. Cho, Materials and Manufacturing Processes **22**, 419–423 (2007).
3.   N. G. Sahoo, Y. C. Jung, N. S. Goo and J. W. Cho, Macromolecular Materials and Engineering **290**, 1049–1055 (2005).
4.   C. Liu, H. Qin and P. T. Mather, Journal of Materials Chemistry **17** (16), 1543-1558 (2007).
5.   J. S. Leng, H. B. Lv, Y. J. Liu and S. Y. Du, Applied Physics Letters **91**, 14410 (2007).
6.   P. R. Buckley, G. H. McKinley, T. S. Wilson, W. Small, W. J. Benett, J. P. Bearinger, M. W. McElfresh and D. J. Maitland, IEEE Transactions on Biomedical Engineering **53**, 2075–2083 (2006).
7.   M. Y. Razzaq, M. Anhalt, L. Frormann and B. Weidenfeller, Materials Science and Engineering: A **444**, 227–235 (2007).
8.   W. Small, IV, P. Singhal, T. S. Wilson and D. J. Maitland, Journal of Materials Chemistry, DOI: 10.1039/b923717h (2010).
9.   T. S. Wilson, J. P. Bearinger, J. L. Herberg, J. E. Marion, III, W. J. Wright, C. L. Evans and D. J. Maitland, Journal of Applied Polymer Science **106** (1), 540-551 (2007).
10.   L.H. Sperling, Introduction to Physical Polymer Science, 4th Ed. (John Wiley & Sons, New Jersey, 2006).
11.   R. J. Seyler, Assignment of the Glass Transition, Volume 1294 of STP/ASTM (ASTM International, 1994) p. 90.
12.   R. J. Seyler, Assignment of the Glass Transition, Volume 1294 of STP/ASTM (ASTM International, 1994) p. 107.

Mater. Res. Soc. Symp. Proc. Vol. 1274 © 2010 Materials Research Society   1274-QQ05-13

Crystallization of Synthetic Hemozoin (Beta-Hematin) Nucleated at the Surface of Synthetic Neutral Lipid Bodies

Anh N. Hoang,[a] Kanyile K. Ncokazi,[b] Katherine A. de Villiers,[†b] , David W. Wright,*[a] Timothy J. Egan*[b]

[a] Department of Chemistry, Vanderbilt University, Station B351822, Nashville, TN 37235, USA. E-mail: David.Wright@vanderbilt.edu

[b] Department of Chemistry, University of Cape Town, Private Bag, Rondebosch 7701, South Africa. E-mail. Timothy.Egan@uct.ac.za

ABSTRACT:

The mechanism of formation of hemozoin, a detoxification by-product of several blood-feeding organisms including malaria parasites, has been a subject of debate; however, recent studies suggest that neutral lipids may serve as a catalyst. In this study, a model system consisting of an emulsion of synthetic lipid bodies, resembling their *in vivo* counterpart in composition and size, was employed to investigate the formation of β-hematin, synthetic hemozoin, at the lipid-water interface. The introduction of heme (Fe(III)PPIX) to this synthetic neutral lipid bodies system under biomimetic conditions (37 °C, pH 4.8) produced beta-hematin with apparent first order kinetics and an average half life of 0.5 min. TEM of monoglycerides (MPG) extruded through a 200 nm filter with heme produced beta-hematin crystals aligned and parallel to the lipid/water interface. TEM data suggests that beta-hematin crystallizes via epitaxial nucleation at the lipid-water interface through interaction of Fe(III)PPIX with the polar head group and elongation occurs parallel this interface.

INTRODUCTION:

During the pathogenic blood stage of a malaria infection, the *Plasmodium* parasites enter the host's red blood cell (RBC) and degrade hemoglobin as a source of nutrition. As a consequence, free heme, known to be capable of causing lipid peroxidation[1] and to destabilize membranes through a colloid osmotic mechanism[2], is released. It is believed that the parasite circumvents heme toxicity by sequestering the heme molecules into a dark brown crystalline material known as the malaria pigment or hemozoin. Hemozoin is a microcrystalline cyclic dimer of ferriprotoporphyrin IX (Fe(III)PPIX) in which the propionate side chain of one protoporphyrin coordinates to the iron(III) centre of the other[3,4]. The dimers hydrogen bond to their neighbours *via* the second propionic acid group, forming extended chains through the crystal. This crystal was first recognized as the discoloration of internal organs of malaria victims in 1717. Despite its early identification, the chemical and structural properties of hemozoin have only been resolved in the last two decades. The molecular mechanism that drives this crystallization remains a topic of debate. However, it is believed that this process is inhibited by chloroquine and other quinoline and related antimalarials[5]. Therefore, understanding the molecular mechanism behind its formation may help in the elucidation of new strategies for the design of antimalarials.

A recent popular hypothesis for hemozoin formation centers on neutral lipid mediation. Specifically, recent electron microscopy imaging shows the close proximity of hemozoin and

---

[†] Present address: Department of Chemistry and Polymer Science, University of Stellenbosch, Stellenbosch, South Africa.

neutral lipid bodies within the digestive food vacuole[6]. In vitro assays revealed that the incubation of these neutral lipids with heme was sufficient to mediate the formation of beta-hematin, synthetic hemozoin[6,7]. Recently, biologically realistic half-lives of about 5 minutes at 37 °C were reported for beta-hematin formation in a lipid-water interface model[8], suggesting that the interface may play an essential role during heme crystallization. The question remains, how does the lipid-water interface facilitate hemozoin formation? In an attempt to address this question, we have examined the organization of lipids and the rate of beta-hematin formation using a lipid-water interface model. Here we show that the lipid forms an emulsion, with beta-hematin formation occurring at the surface of the synthetic neutral lipid bodies. We also provide a kinetic study of the process of beta-hematin formation in this system. A detailed understanding of beta-hematin formation using the lipid-water interface system may provide more insight into the mechanism of hemozoin nucleation and growth processes in nature.

EXPERIMENTAL METHODS:

The following lipids were used in this study: monoglyceride lipids *rac* 1-monopalmitoylglycerol (MPG), *rac* 1-monostearoylglycerol (MSG), *rac* 1-monomyristoylglycerol (MMG) and *rac* 1-monooleoylglycerol (MOG); diglycerides 1,3-dioleoylglycerol (DOG), 1,3-dipalmitoylglycerol (DPG) and 1,3-dilinoleoylglycerol (DLG); triglycerides trioleoylglycerol (TOG) and trimyristoylglycerol (TMG). A stock 3.3 mM solutions of lipids were prepared in 1:9 acetone/ methanol solvent. Stock solutions made from citric acid were used to prepare citrate buffer (50 mM) and pH adjusted to 4.8 using a slurry of sodium hydroxide (NaOH). A stock solution (3.2 mM) of hematin was formed by dissolving 2 mg of hematin in 0.4 mL NaOH (0.1 M) followed by the addition of 0.6 mL of 1:9 acetone/methanol(from here on generically referred to as heme).

Characterization of synthetic neutral lipid bodies

200 µL of stock MPG solution was layered on top of a 5 mL solution (in 15mL Falcon tube) of citrate buffer (37 °C) drop-wise using a microsyringe. After 5 min. incubation, 2 µL of a 500 µg/mL acetone solution of a hydrophobic phenoxazone dye, Nile Red was introduced to the surface. Samples were then withdrawn from beneath the surface and imaged by a Nikon Epifluorescent microscope and a Zeiss LSM 510 meta inverted confocal microscope. Confocal imaging was performed using a 488 nm argon laser with 505-550 nm band pass filter. Confocal orthogonal projection was produced using Zeiss LSM Image Browser software. An additional sample was stained with a 2 % solution of uranyl acetate for imaging using a JEM 1200EXII TEM operated at 120 kV. The size distribution of the lipid emulsions were measured using two Malvern particle sizing systems: The Malvern Zetasizer with a lower limit of 0.1 nm and an upper limit of 6 µm and the Malvern Mastersizer with a lower limit of 50 nm and upper limit of 800 µm. 1 mL of MPG stock solution was deposited onto surface of citric acid buffer (50 mL, 50 mM, pH 4.8) and 0.5 mL samples extracted from below the surface were used for Zetasizer measurements while 2 mL samples were used for the Mastersizer. Three measurements were taken for each sample at 0, 20, 40, and 60 minutes after the lipids were deposited to determine the stability of the size distribution of the lipid particles in these emulsions with time.

Localization of heme and kinetics of beta-hematin formation

2 µL solution of stock heme and 200 µL of stock lipids (MPG) were premixed and was deposited onto the surface of citric acid buffer solution (5 mL, 37 °C, pH 4.8). To determine the localization of heme with respect to the lipid bodies, samples were immediately taken from

beneath the layered surface and prepared for TEM imaging using a LEO 912 transmission electron microscope. Additional samples were withdrawn after ten minutes of incubation at 37 °C. Iron distribution was measured from the latter sample on objects that resembled beta-hematin crystals using TEM electron spectroscopic imaging (ESI) for elemental analysis of iron. For kinetic studies, the same ratio of lipid-heme premix was deposited onto buffer and incubated at 37 °C for 0, 1, 3, 5, 15, 30, 45 and 60 minutes. The reaction was quenched with the addition of 1 mL of an aqueous solution containing 30% pyridine, 10% HEPES buffer (2.0 M, pH 7.5), and 40% acetone (v/v).[9] The sample was immediately centrifuged for 10 minutes at 4,000 rpm. The percentage of unconverted heme was determined by performing colorimetric measurements at 405 nm on the supernatant of quenched and centrifuged samples.

Formation and characterization of beta-hematin

Product formed in the lipid-water system was characterized by Fourier transformed infrared (FTIR) spectroscopy. XRD was carried out using Cu Kα radiation (λ=1.541 Å), with data collection on a Philips PW1050/80 vertical goniometer in the 2θ range 5 – 40° using an Al sample holder. TEM images were obtained using the LEO 912. Heme incubated in the aqueous medium alone served as control.

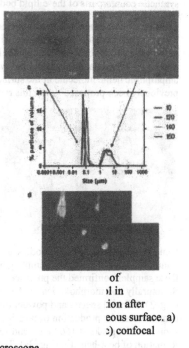

Beta-hematin formed via extrusion method

A premixed solution of MPG (1 mL, 3.3mM) and hematin (10 μL, 3.2mM)) was adjusted to pH 7.0 with 0.01 M hydrochloric acid (HCl). Citric acid buffer (~300 μL, 50 mM, pH 7.0) was carefully added to this solution, to avoid precipitation of the lipid solution. MSG/heme vesicles were formed by extrusion through Nucleopore track-etch polycarbonate membranes (Avanti Polar Lipids Inc.) 19 times at temperature above the T_c. The newly formed heme-MSG lipid emulsions were introduced into an aqueous solution buffered at pH 4.8 containing 50 mM citrate. The mixture was incubated at 37 °C for about 10 minutes and TEM imaging of 3 μL of samples was taken using the LEO 912 electron microscope.

of
l in
ion after
eous surface. a)
c) confocal

RESULTS AND DISCUSSION:

Characterization of synthetic neutral lipid bodies

To investigate the organization of neutral lipids when deposited onto an aqueous buffer, transmission electron microscopy (TEM) and fluorescent imaging were performed on samples extracted from just beneath the surface. Imaging data showed that lipid molecules spontaneously organized into a lipid emulsion beneath the surface (Fig. 1a, b). Fluorescence imaging of samples stained with the lipophilic stain, Nile Red, and a water soluble dye, fluorescein, showed lipid bodies in the emulsion layer in the micrometer size range (Fig. 1b), while TEM imaging showed structures in the nanometre size range (Fig. 1a). Confocal microscopy of the samples confirmed that the lipid bodies contain a homogeneous core (Fig. 1d) and are not hollow liposomes. Measurements of the lipid bodies size distribution in the lipid emulsions revealed two populations. A large population in the 50 to 200 nm diameter range (Fig. 1c) and a smaller population of synthetic lipid bodies in the 1 to 10 μm range. These emulsions were stable over time, as little fluctuation in size distribution was observed in four independent measurements over a period of an hour. These lipid bodies, which form spontaneously are strikingly similar in appearance to neutral lipid bodies reported associated with the digestive vacuole of *P. falciparum* and the gut lumen of *S. mansoni* and which, at least in the case of the former organism have been shown to consist largely of mono-, di- and triglycerides[6,7]. Indeed, they appear to be essentially synthetic counterparts of these lipid bodies.

Localization of heme and product characterization

When hematin was premixed with MPG and deposited onto an aqueous surface, electron dense material, probably heme, rapidly localized around the surface of the synthetic neutral lipid bodies, apparently at or near the lipid-water interface (Fig. 2a). Such accumulation of the amphiphilic heme molecule at the surface of lipid bodies in *P. falciparum* and *S. mansoni* has previously been proposed[6,7], but this represents the first direct evidence of localization of free

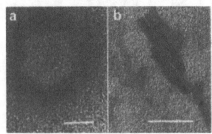

Fig 2. TEM showing heme localization at or near the lipid-water interface. When haematin was premixed with MPG and introduced onto an aqueous surface and immediately imaged using TEM, electron dense material, presumably hematin, was localized at or near the lipid-water interface (a). Elemental imaging of iron using TEM with electron spectroscopic imaging (ESI) conducted on samples reacted for 10 min. confirmed the presence of beta-hematin crystals (b). Regions highlighted in red denote the presence of Fe. Scale bar = 50 nm.

heme at the interface of lipid bodies and water. TEM images of samples incubated for 10 minutes displayed crystals resembling beta-hematin (Fig. 2b). TEM/ESI of iron distribution in these samples confirmed the presence of iron in these putative beta-hematin crystals (Fig. 2b). Structurally and morphologically, the crystals formed resemble hemozoin in their crystal habit (Fig. 3). Infrared spectra and powder diffraction data of the products formed using MPG clearly demonstrates the production of beta-hematin with percentage of product conversion of 65 ± 4. Prominent IR peaks at 1662 cm^{-1} and 1210 cm^{-1} are readily discernible indicating significant formation of beta-hematin (Fig. 3a). These findings are confirmed by XRD (Fig. 3b) showing the characteristic diffraction peaks of beta-hematin.

Kinetic studies of beta-hematin formation

A premixed lipid-hematin solution was deposited onto the aqueous sub-phase and incubated at designated time. After 60 minutes, the percent yields were similar for mono- and diglycerides while less product was formed using triglycerides. Overall, the reaction was completed within five minutes with a significant quantity of product was formed in less than one minute (Fig. 4a). If hemoglobin degradation occurs at a constant rate, lipid-water interface mediated hemozoin formation would be more than sufficient to manage the entire free heme flux during the intraerythrocytic stage.

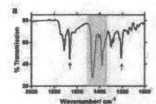 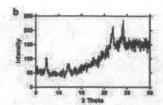 

Fig. 3. Infrared (a), X-ray diffraction (b), and transmission electron microscopy (c) evidence for the formation of β-hematin at the lipid-water interface using MPG. Two large Nujol peaks (a) obscure the region between 1550 and 1320 cm^{-1} but do not interfere with the characteristic β-hematin IR peaks at 1660 and 1210 cm^{-1}. Scale bar = 50nm.

Unsaturated glycerides promoted heme conversion more rapidly than saturated glycerides. Nearly a two fold increase in rate was observed for formation of beta-hematin under unsaturated glyceride (MOG, DOG, and TOG) mediation when compared to the rate calculated from saturated glycerides (Fig 4). In the absence of the synthetic lipid bodies, no product was formed. The synthetic neutral lipid bodies system presented here is competent for the detoxification of free heme at a biologically realistic rate under physiological conditions. Indeed, the observed rates are the fastest so far reported for beta-hematin formation with half-lives of about 0.5 min at 37 °C for the monoglycerides MPG, MMG and MSG and only slightly slower for the di- and triglycerides (DLG, DPG and TMG). Thus, the neutral lipid body-water interface is likely to be the site of hemozoin formation *in vivo*.

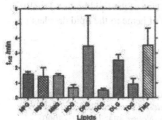

Fig 4. A) Rate of beta-hematin formation under the mediation of various neutral lipids.

Beta-hematin formation via methods of extrusion

TEM imaging of extruded samples of MSG/heme solutions showed beta-hematin formed near the outer region of the synthetic lipid bodies (Fig. 5). In all observations, crystal size did not exceed dimensions of associated lipid bodies. Clusters of beta-hematin crystals were observed localized near and aligned parallel to the interface and each other. In several images, the beta-hematin crystals were located along the lipid-water interface of the synthetic lipid bodies. These images closely resemble observations in *P. falciparum* previously reported[6,10], while images of early stages of hemozoin formation in *S. mansoni* also appear to be similar upon close inspection[11]. The observations in the current study provide direct evidence that crystal nucleation occurs at the lipid-water interface and that crystal elongation occurs parallel to this interface. Based on the previous work which relates the shape of beta-hematin crystals to the {h,k,l} indices of the crystal, it appears likely that crystal extension occurs parallel to the c-axis until its growth is limited by the curvature of the lipid bodies[12]. Thus, the slowest growing {100} face is in contact with the lipid surface and the fastest growing {001} face is approximately perpendicular to the interface. However, in the absence of electron diffraction data in the present study, this remains somewhat speculative.

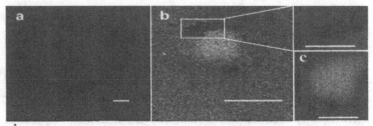

Fig 5. TEM of beta-hematin formed at the MSG/water. Crystals were observed aligned along the lipid-water interface for both extruded samples of MSG. Clusters of β-haematin crystals were oriented parallel to each other and to the lipid-water interface. Scale= 50nm

CONCLUSION:

Experimental evidence supports the hypothesis that the nucleation and extension of beta-hematin is localized at the interface of neutral lipid bodies and water and that these bodies are responsible for the reproducibility of the structural morphology and size of hemozoin crystals observed in nature. This model accounts for the orientation and formation of beta-hematin crystals with respect to these bodies. In particular, the model stems from the observation made during these experiments, which revealed the presence of (i) heme localization and nucleation at the lipid-water interface, (ii) crystal extension along the interface, (iii) orientation of mature crystals and (iv), crystal sizes that appear to be limited by the size of the lipid bodies. Finally, it should be noted that it is likely that the transport and delivery of heme to the lipid droplets *in vivo* will differ in several respects.

REFERENCES:
(1)S.H. Vincent, *Semin. Hematol.* **26**, 105-11(1989).
(2)A.C. Chou, C.D. Fitch, *J. Clin. Invest.* **68**, 672-677 (1981).
(3)A. F. Slater, W.J. Swiggard, B.R. Orton, W.D. Flitter, D.E. Goldberg, A. Cerami, G.B. Henderson, *Proceedings of the National Academy of Sciences of the United States of America* **88**, 325-329 (1991).
(4)S. Pagola, P.W. Stephens, D.S. Bohle, A.D. Kosar, S.K. Madsen, *Nature* **404**, 307-310 (2000).
(5)J. Ziegler,R. Linck, D.W. Wright, *Curr Med Chem.* **8**, 171-89 (2001).
(6) J.M. Pisciotta,I. Coppens, A. K. Tripathi, P.F. Scholl, J. Shuman, S. Bajad, V. Shulaev, D. J. Sullivan, *Biochem J.* **402**, 197-204 (2007).
(7)K. E. Jackson, N. Klonis; D.J.P. Ferguson, A. Adisa, C. Dogovski, L. Tilley, *Mol Microbiol.***54**, 109-122 (2004).
(8) T.J. Egan, J. Y.-J. Chen, K. A. de Villiers, T. E. Mobotha, K. J. Naidoo, K.K.Ncokazi, S.J. Langford, D. McNaughton, S. Pandiancherri, B.R. Wood, *FEBS Lett.* **580**, 5105-5110 (2006).
(9) K.K. Ncokazi, T.J. Egan, *Anal Biochem.***338**, 306-319 (2005).
(10) I. Coppens, O. Vielemeyer, *International Journal for Parasitology* **35**, 597-615 (2005).
(11) M.F. Oliveira, J.C.P. d'Avila, C.R. Torres, P.L. Oliveira, A.J. Tempone, F.D. Rumjanek, C.M.S. Braga, J.R. Silva, M. Dansa-Petretski, M.A. Oliveira, W. de Souza, S. T. Ferreira, *Molecular and Biochemical Parasitology* **111**, 217-221 (2000).
(12)R. Buller, M.L Peterson, O. Almarsson, L. Leiserowitz, *Cryst. Growth Des.* **2**, 553-562 (2002).

Mater. Res. Soc. Symp. Proc. Vol. 1274 © 2010 Materials Research Society 1274-QQ05-17

The Effect of Type 1 Diabetes on the Structure and Function of Fibrillin Microfibrils

R Akhtar[1,2], JK Cruickshank[2], NJ Gardiner[3], B Derby[1], MJ Sherratt[4]
[1] School of Materials, Grosvenor Street, The University of Manchester, Manchester, M1 7HS, United Kingdom.
[2] Cardiovascular Sciences Research Group, Manchester BioMedical and Academic Health Sciences Centre, The University of Manchester, 46 Grafton St, Manchester, M13 9NT, United Kingdom.
[3] Faculty of Life Sciences, AV Hill Building, Oxford Road, The University of Manchester, Manchester, M13 9PT, United Kingdom.
[4] School of Biomedicine, Stopford Building, The University of Manchester, Oxford Road, Manchester, M13 9PT, United Kingdom.

ABSTRACT

In both Type 1 and 2 diabetes tissue stiffening is evident from measurements of the gross mechanical properties of the vasculature. Whilst composite elastic fibres play an important role in mediating vascular elasticity in healthy individuals, the effects of diabetes on the structure and function of individual elastic fibre components remains poorly defined. Fibrillin microfibrils, which are key structural elements of the elastic fibre system, have a unique 'beads-on-a-string' morphology and a mean periodicity of approximately 56 nm. In this study, we have characterised the effects of experimentally induced Type I diabetes on the structure of microfibrils isolated from rat aorta. Microfibril length and periodicity were quantified from atomic force microscopy (AFM) images. Although there was no significant difference in mean microfibril length between healthy and diabetic animals (control 23.2 repeats, SEM 6.2 repeats: diabetic 23.6 repeats, SEM 6.1 repeats: Mann Whitney U-test, p=0.391), mean periodicity was significantly reduced in microfibrils isolated from the diabetic rats (52.7 nm, SEM 0.4 nm) compared with age-matched controls (59.5 nm, SEM 0.4 nm) (p<0.001, Student's t-test). In conclusion, this study demonstrates that key elastic fibre components undergo significant structural alterations in diabetic individuals.

INTRODUCTION

Arterial stiffening which is a feature of Type 2 diabetes is also evident in Type 1 diabetes irrespective of age [1]. Whilst the pathological glycosylation of extracellular matrix proteins is thought to play an important role in increasing tissue stiffness in diabetic patients, the molecular mechanisms through which diabetes leads to increased cardiovascular risk are not clearly defined [2].

Elastic fibres are major structural and functional components of dynamic tissues. In large arteries, such as the aorta, elastic fibre-containing lamellae store energy under physiological loads to drive recoil back to the resting state, thus allowing these vessels to withstand high blood pressure. Currently however, the effects of diabetes on individual components of the elastic fibre system are poorly understood. Fibrillin microfibrils, which are an important component of the elastic fibre system, are thought to play an important mechanical role, acting as a stiff reinforcing phase within a highly compliant elastin matrix [3]. Fibrillin microfibrils have a 'beads-on-a-string' structure with a periodicity (bead-to-bead distance) of approximately 56 nm (Figure 1).

Figure 1. Fibrillin microfibril extracted from rat aorta and imaged by AFM.

Given the key-role played by fibrillin microfibrils in mediating tissue elasticity and the profound alterations in tissue mechanical properties that accompany prolonged diabetes, we hypothesized that fibrillin microfibril structure would be significantly compromised in hyperglycaemic (diabetic) conditions. In this study, therefore we have characterized the structural changes induced in fibrillin microfibrils isolated from rats with experimentally induced Type 1 diabetes [4].

EXPERIMENT

Diabetes was induced in adult male Wistar rats (250-300g) with an intraperitoneal injection of streptozotocin (STZ; 55 mg/kg, n=3). Hyperglycaemia was confirmed (>15 mmol/l) three days following the STZ injection. Age-matched control rats acted as controls (n=3). Rats were sacrificed 8 weeks post-STZ. Fibrillin microfibrils were isolated and purified from the descending aorta by bacterial collagenase digestion and size-exclusion chromatography using well-established methods [3].

AFM Sample Preparation

AFM samples were prepared by adhering mica to 15 mm diameter metal sample stubs using clear nail varnish. Negatively charged, highly hydrophilic mica was freshly cleaved and then coated with positively charged poly-L-lysine (PLL). By reducing surface charge, this chemical modification promotes the retention of native microfibril morphology following adsorption. Aliquots containing isolated microfibrils were diluted 1:5 in buffer solution (400 mM NaCl, 50 mM, Tris–HCl, pH 7.4). The diluted suspension was pipetted on to the mica-PLL surface and incubated for 1 minute following which excess liquid was removed with filter paper. Fibrillin microfibrils are stabilized by salt bridges [6] and therefore the samples were allowed to dry in the salt-rich buffer prior to removal of crystallized salt by washing the surface with 1 ml of distilled water and further drying in air.

Image Acquisition and Analysis

AFM imaging was conducted in tapping mode in air using a Multimode AFM with a Nanoscope IIIa controller (Veeco Instruments Ltd, Cambridgeshire, UK) using silicon nitride probes with a nominal resonant frequency of 300 kHz at a scan speed of 1.49 Hz (OTESPA probes, Veeco Instruments Ltd., Cambridgeshire, UK). AFM height images of microfibrils (2μm window, 512 × 512 pixels) were collected for length and periodicity analysis.

Axial height profiles (Figure 2) were extracted using WSxM 3.0 (Nanotec Electronica, S.L, Spain) [5] and exported in ASCII format to OriginLab (Northampton, MA, USA). Microfibril periodicity was determined using the Peak Analyzer tool within OriginLab. In addition, microfibril length was quantified by counting the number of repeats per microfibril.

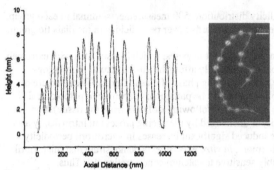

Figure 2. Typical axial height profile for a 20 bead fibrillin microfibril extracted from the aorta of a diabetic rat.

DISCUSSION

Abundant microfibrils were isolated from both control and diabetic rat aortas and there was no significant difference in mean microfibril length (control 23.2 repeats, SEM 6.2 repeats: diabetic 23.6 repeats, SEM 6.1 repeats: Mann Whitney U-test, p=0.391). This observation suggests that fragmentation of this elastic fibre component may not play a role in mediating the altered elasticity which characterizes many diabetic tissues. In contrast, mean periodicity was significantly reduced in microfibrils isolated from diabetic rats (52.7 nm, SEM 0.4 nm) compared with age-matched controls (59.5nm, SEM 0.4 nm) (p<0.001, Student's t-test) (Figure 3).

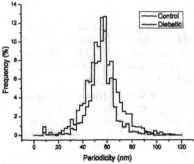

Figure 3. Fibrillin microfibril periodicity distribution (500 measurements/animal in each group; 1500 measurements per group). Note the shift towards lower periodicities in the diabetic group.

Therefore, even short-term insulin-deficient diabetes, which is associated with extremely high blood glucose levels, is sufficient to profoundly affect the ultrastructural morphology function of fibrillin microfibrils (the key reinforcing phase of elastic fibres). The structure of these important extracellular assemblies appears to be particularly sensitive to pathological changes both *in vivo* (within tissues) and *in vitro* (following isolation from the tissue). For example, whilst microfibril periodicity is unchanged by the ageing process in intrinsically aged skin [7], exposure to cigarette smoke induced significant increases in microfibril periodicity in the skin of heavy smokers [8]. Furthermore, *in vitro* experiments demonstrate that microfibril periodicity is exquisitely and reversibly sensitive to calcium concentration [9]. Thus, environmental factors may have a profound effect on fibrillin microfibril structure and hence function.

Of particular interest is the relationship between ultrastructural morphological changes in elastic fibre components as characterized in this study and the overall mechanical properties of the aorta. The gross mechanical properties of diabetic rat aortae undergo significant changes compared to healthy controls. The Young's modulus is increased whilst maximum strain is decreased 7 weeks after STZ injection [10]. Thus, pathological changes in gross mechanical properties have been reported in the same diabetes model which was employed in this study. However, the molecular mechanisms which are responsible for the morphological changes in isolated microfibrils remain to be determined. Further work is required to determine if these changes in microfibril structure are a cause or an effect of the pathological changes observed in the tissue.

CONCLUSIONS

In an extreme diabetic environment, there were significant effects on the structure of vascular fibrillin microfibrils.

ACKNOWLEDGMENTS

The authors would like to thank British Heart Foundation and Research into Ageing for funding this work.

REFERENCES

1. N. Winer, J. Sowers, *"Atherosclerosis, Large Arteries and Cardiovascular Risk.* Advances in Cardiology", Diabetes and Arterial Stiffening, ed. M.E. Safar and E.D. Frohlich, (Karger, 2007) pp 245.
2. M.T. Schram, R.M.A. Henry, R.A.J.M. van Dijk, P.J. Kostense, J.M. Dekker, G. Nijpels, R.J. Heine, L.M. Bouter, N. Westerhof, C.D.A. Stehouwer, Hypertension, **43**, 176 (2004).
3. M.J. Sherratt, C. Baldock J.L. Haston, D.F. Holmes, C.J.P. Jones, C.A. Shuttleworth, T.J. Wess, and C.M. Kielty, J. Mol. Biol. **332**:183 (2003).
4. H. Bar-On, P.S. Roheim, and H.E. Eder, Diabetes, **25**, 509 (1976).
5. M.J. Sherratt, T.J. Wess, C. Baldock, J.L. Ashworth, P.P. Purslow, C.A. Shuttleworth, and C.M. Kielty, Micron, **32**, 185 (2001).
6. I. Horcas, R. Fernandez, J.M.Gomez-Rodriguez, J. Colchero, J. Gomez-Herrero,A.M. Baro, Rev. Sci. Instrum., **78**, 013705-1 (2007).
7. M.J. Sherratt J.Y. Bastrilles J.J. Bowden, R.E.B. Watson, C.E.M. Griffiths, Brit. J. Dermatol. **155**, 240 (2006).
8. H.K. Merrick, C.E.M. Griffiths, E. Tsoureli-Nikita, R.E.B. Watson and M.J. Sherratt, Brit. J. Dermatol. **160**, 916 (2009).
9. T.J. Wess, P.P. Purslow, M.J. Sherratt, J. Ashworth, C.A. Shuttleworth, and C.M. Kielty, J. Cell. Biol. **141**, 829 (1998).
10. G.K. Reddy, Microvasc. Res., **68**, 132 (2004).

Nanomechanics of Biological Molecules and Structures

Mater. Res. Soc. Symp. Proc. Vol. 1274 © 2010 Materials Research Society   1274-QQ06-02

A Strategy to Simulate the Dynamics of Molecular Assemblies Over Long Times

Julius T. Su[1]
[1]Materials and Process Simulation Center, Caltech, Pasadena, CA 91125, U.S.A.

ABSTRACT

We propose a strategy to simulate the dynamics of molecular assemblies over long times, provided they have a hierarchical and modular nature. In the scheme, fast fluctuations are averaged into a set of effective potentials (fluctuation softened potentials or FSPs), and the remaining slower dynamics are propagated in a drastically reduced configuration space (coupled energy landscapes or CEL). As a preliminary validation of the FSPs we compute the free energy of binding of a protein complex (RNase:barstar) for different relative positionings of the proteins. As a demonstration of CEL, we simulate the dynamics of microtubule unraveling upon hydrolysis of bound nucleotides. The method should allow the use of time steps hundreds to thousands of times longer than in conventional molecular dynamics, so that with only atomic structures and interactions as input, motions over human time scales ($>ms$) could be simulated.

INTRODUCTION

In all living organisms, key cellular processes are performed and regulated by assemblies of macromolecules. For example, proteins in *E. coli* bacteria are synthesized by their ribosomes, 82 proteins enmeshed in 3 RNA molecules that collectively form a molecular machine. Understanding better how such assemblies function and misfunction is a critical step toward developing new and effective therapeutics for a broad range of human diseases.

Computation has long held out the tantalizing prospect of being able to simulate biological function starting from atomic structures and interactions, via a suitable series of approximations and step-by-step dynamics. As yet, however, such approaches have been stymied by the huge gap between the fast time scales of atomic motions (10^{-15} seconds), and the relatively slow time scale of interesting biological motions (10^{-3} seconds and beyond) – simply too many intermediate computations need to be performed. This difficulty has been colloquially termed the "time scale problem."

In this paper, we do not propose a solution for the time scale problem in general, but suggest instead that *for a certain class of systems*, vast accelerations in the speed of dynamics simulations ought to be possible, making simulations over biological time scales practical.

We focus on simulating molecular assemblies with the following two characteristics: (1) hierarchy, i.e. there is a nesting of components within components over a range of time and length scales, and (2) modularity, i.e. the components of the assembly retain characteristics of their individual dynamics, even in the context of the complex.

These properties may be exploited to make long time dynamics simulations feasible as follows. First, hierarchy imposes a nested separation of time scales. We pick the time scale of protein conformational transitions ($> \mu s$) as a dividing point. Fluctuations faster than that are averaged out into effective potentials we term *fluctuation softened potentials*, while motions slower than that are explicitly computed using a scheme we term *coupled energy landscape* dynamics. This method allows the use of long time steps in the propagation of dynamics, making long duration simulations possible.

Second, modularity implies that inter-component interactions perturb, but do not overwhelm, the dynamics of individual components. From this principle, we derive a strategy we term "watch and combine", where we sample (watch) the motions of individual components, then use information extracted from the sampling to aid in computing properties of the assembly.

There is mounting evidence from experiment and theory that the dynamics of many biological systems are both hierarchical and modular. In support of the first property, combined data from NMR, FRET, X-ray, and MD methods link motions on a hierarchy of time scales to enzymatic activity in the adenylate kinase protein – i.e. fast side chain motions (ps) drive slower backbone motions (ns), which drive still slower domain motions (μs to ms), which affect catalysis [1]. In support of the second property, isolated proteins in solution execute motions similar to those made when they bind in complexes, c.f. an NMR study of ubiquitin [2]. If these properties hold true in a broad way, the methods presented here would be applicable to study a wide range of biological phenomena.

This paper is organized as follows. First, we outline the fluctuation softened potential method, used to average free energies over nanoseconds; and its application to study the binding of two proteins (RNase and barstar). Second, we outline the coupled energy landscape method, used to compute dynamics of assemblies over microseconds and beyond; and its application to study the unraveling of microtubules (which occurs upon hydrolysis of bound nucleotides). Finally, we conclude by comparing the methods presented here to other methods for long time simulations, and discuss what we are likely to learn from the current endeavor.

THEORY

Fluctuation softened potentials for estimating free energies averaged over nanoseconds

We introduce a method for estimating quickly the free energy of interaction between multiple elements of an assembly, averaging over conformational fluctuations of the components, and also their rigid body rotations. The overall procedure is outlined in Figure 1.

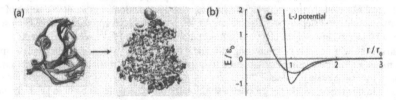

Figure 1. Outline of the fluctuation softened potential method for estimating the binding free energy between multiple proteins. A "watch and combine" strategy is used: (a) For each protein, fluctuations of individual atoms – anisotropic B-factors – are obtained from molecular dynamics simulations, elastic network models, or experiment. (b) The interaction potentials between proteins are modified by atomic fluctuations to generate free energy interaction potentials which when summed provide an estimate of the overall binding free energy.

As a starting point, we assume we possess an implicit solvent force field that given an instantaneous snapshot of the position of all atoms in the system, returns an overall interaction energy. We write the free energy integral as follows:

$$e^{-\Delta G/kT} = \int e^{-\Delta E(\{r_i\})/kT}\,dr_i = \int\!\left(\int e^{-\Delta E(\{r_{conf},\,r_{rigid},\,\Omega_{rigid}\})/kT}\,dr_{conf}\right)dr_{rigid}\,d\Omega_{rigid}$$

To evaluate the integral over conformation space, we make a severe approximation – we assume that atoms on separate proteins, taken pairwise, have motions that are completely uncorrelated. Then the integral reduces to a sum of free energy potentials, which are based on the interaction energy function between atoms, modified by the probability density function of fluctuations:

$$G[\rho_i,\rho_j] = -kT\log\int e^{-E(r_{12})/kT}\cdot \rho_i(r_1)\rho_j(r_2)\,dr_1\,dr_2$$

These probability density functions – one for each atom – may approximated by any convenient function – we use Gaussian ellipsoids, specified by six parameters each, i.e. anisotropic B-factors. The resulting parameterized free energy functions we term "fluctuation softened potentials" (or FSPs). Given a set of FSPs, evaluating the overall free energy is straightforward – it becomes a pairwise sum, sampled and averaged all accessible translations and orientations of the components in the complex. This sampling may be accomplished using any convenient Monte Carlo or dynamics scheme.

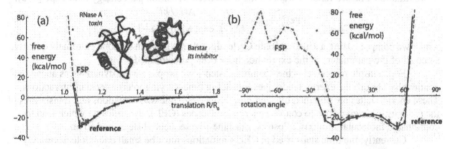

Figure 2. Validation of the fluctuation softened potential method, for van der Waals interactions only. We compute free energies for a protein complex (RNase A:barstar) for different relative (a) translations and (b) rotations of the proteins, comparing values from the FSP (red dashed curve) to values from a reference ensemble method (black solid curve, gray dots are free energies from a random subdivision of the conformer pairs into ten bins – their spread provides an estimate of error). The agreement is excellent, with no use of empirical parameters.

To validate the FSP method (Figure 2), we computed the binding free energy of a protein complex (barstar:RNase A) for different relative positionings of the constituent proteins, comparing FSP to a reference theory – free energies computed from a 2500 by 2500 conformer ensemble cross-sampled over 1 ns. The agreement is excellent, with no use of empirical parameters.

As an energy function, we used an implicit solvent force field previously developed to study the association of flexible proteins ([3], AMBER/DDD/SASA). Currently, we have only parameterized a FSP for van der Waals (vdW) interactions, so we compared only the free energy contributions from the vDW component of energy. FSPs that compute free energy contributions from electrostatic and solvation interactions are under development.

Coupled energy landscapes for propagating dynamics beyond microseconds

We outline a procedure for computing the major conformational dynamics of macromolecules, and estimating how those dynamics are modified by time-varying interactions between molecules (Figure 3).

We use a "watch and combine" strategy. In the "watch" step, each protein is considered separately, and transition paths connecting major conformations are mapped out, using a method such as nudged elastic band theory. The transition paths are then broken down into a set of discrete points, each containing an ensemble of conformations – and their fluctuation parameters – as well as a free energy and an effective diffusion constant to transit from one point to another.

We make two key assumptions, based on the presence of hierarchical dynamics. First, we assume that conformations change slowly along the transition path, so that in hopping from one point to another, all memory of previous motion is lost. Second, we assume all other non-transition path motions are fast, so that they are thoroughly sampled and averaged over at each step.

If these assumptions hold, the overall dynamics of an individual molecule is to a good approximation a biased random walk on the discretized transition path/energy landscape [4, 5]:

$$\text{rate}(i \rightarrow j) = \frac{D}{(\Delta x_{ij})^2} \frac{\Delta G_{ij}/kT}{\exp(\Delta G_{ij}/kT)-1}$$

which we compute using a kinetic Monte Carlo algorithm – this can be done extremely quickly, since all of the parameters in the expression have been precomputed.

For multiple molecules – the "combine" step – we propagate the dynamics as above, but continually perturb the energy landscapes by adding time-varying intermolecular interactions. These energy deltas are obtained by averaging FSPs over rigid body rotations and translations of the proteins. This gives rise to coupled energy landscapes (CEL), dynamics on which lead to coordinated molecular motions, allostery, machine-like sequential steps, and so on.

Currently, the time steps used in CEL simulations must be small enough to accurately capture rigid body rotations and translations between proteins interacting via soft potentials (FSPs). In practice, using Brownian dynamics, this works out to $\Delta t \sim 0.1$ to 1 ps, i.e. roughly a hundred to a thousand times longer than time steps for conventional molecular dynamics.

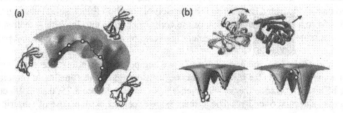

Figure 3. Outline of the coupled energy landscape method for simulating the dynamics of molecular assemblies over long times. (a) Individual proteins are represented by transition paths between major conformations, broken down into a set of discrete points. This forms a free energy landscape upon which coarse dynamics can be propagated. (b) When multiple proteins interact, their prestored energy landscapes are continually modified by intermolecular interactions, giving rise to higher order coordinated motions.

As an application of CEL, we simulated the dynamics of microtubule unraveling upon nucleotide hydrolysis (Figure 4). Microtubules are key elements of the cytoskeleton, and when cells divide, they must be disassembled and subsequently reassembled in a controlled fashion – indeed, many anticancer drugs (i.e. paclitaxel and *Vinca* alkaloids) act by targeting microtubules. Yet key questions remain concerning the microscopic details of microtubule assembly and disassembly [6, 7]. We are working to develop a better understanding of this process.

Microtubules are assemblies of tubulin dimers, each of which has bound a nucleotide (GTP, which when hydrolyzed becomes GDP). One model suggests that microtubule unraveling is driven by changes in the geometry of tubulin upon hydrolysis of its bound nucleotide – it is estimated that GTP-tubulin has a preferred bending angle of 5°, while GDP-tubulin has a preferred bending angle of 12° (from [7]); GDP-tubulin with its higher bending angle is less able to fit into a regular lattice, which causes the tubular structure to peel apart.

To test this model, we created a coarse-grained (CG) representation of tubulin dimer as 8 beads in a wedge-like arrangement, with charges at the top and bottom of the wedge making it possible to form strong lateral contacts with adjacent dimers (VDW interactions form weaker longitudinal contacts). As energy landscapes, we used double well potentials, since it is known there is a barrier to interconversion between straight and bent tubulin forms [8]; we then biased the GTP and GDP-tubulin landscapes toward straight and bent structures respectively.

In the simulations, we observed that microtubules made of GDP-tubulin were stable when capped by GTP-tubulin, as in nature, but unraveled when the caps were hydrolyzed to GDP-tubulin, also observed experimentally [9]. Clearly, this simulation serves as a proof-of-concept only – all aspects of the coarse-grained particles and potentials were only estimates based on "reasonable" experimental quantities, i.e. [10]. The ultimate goal is to perform CEL simulations with all parameters derived from atomistic simulations only, using path sampling methods and the FSP methodology discussed previously.

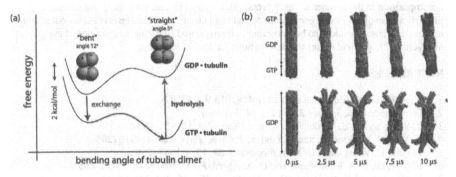

Figure 4. Dynamics of microtubule unraveling simulated using the coupled energy landscape method. (a) Tubulin dimer is represented by coarse grained beads and simple double well energy landscapes; (b) the CEL simulation shows that GTP-capped GDP-microtubules are stable, while GDP-only microtubules unravel over 10 microseconds.

CONCLUSIONS

The method we propose to enable long time simulations is a simple extension of an old idea: the explicit propagation of slow "functional" motions, together with the implicit averaging of fast "nonfunctional" motions. Indeed, this separation is the basis of all CG methods.

We go further by focusing our attention on molecular assemblies where strong assumptions of modularity and hierarchy apply. Modularity makes it possible to precalculate quantities that are usually computed on the fly, or assumed constant – such as the contribution atomic fluctuations make to the free energy of interacting pairs of atoms. Hierarchy imposes a separation of time scales, so that stochastic dynamics on a time-varying free energy landscape may be used as a surrogate for full atomistic dynamics.

With the proposed method one could precompute the dynamics of a whole set of components, then quickly evaluate the dynamics of every conceivable "mixed-and-matched" arrangement of them. This would be a boon to design.

Within the proposed scope of applicability, the combined FSP/CEL method has the potential to be faster and more accurate than other methods that do not make assumptions about hierarchy and modularity, e.g. subspace dynamics, where dynamics is propagated along the lowest order normal modes only [11], relies on an on-the-fly separation of time scales because hierarchy is not assumed; and coarse-grained beads [12] and softened potentials for dissipative particle dynamics [13] rely on potentials that are distinguished by atom type, not by environment – "watch and combine" strategies are not used because modularity is not assumed.

An immediate question is whether the promised gains in accuracy and speed for this method will materialize. To some extent this can be achieved simply by picking the "correct" model systems to study. But a broader and more philosophically interesting question is whether there exists a large range of systems, biological and otherwise, whose dynamics are amenable to decomposition in the manner we have proposed. If hierarchy and modularity are universal properties of evolved or designed systems, that is an observation that deserves to be noted and exploited. If not, it would also be intriguing to discover and study the seams where these abstractions "leak", and report the observations for future reference.

REFERENCES

1. K. A. Henzler-Wildman et al, *Nature* **450**, 913-916 (2007).
2. O. F. Lange et al, *Science* **320**, 1471-1475 (2008).
3. T. Wang and R. C. Wade, *Prot: Struct. Func. Gen.* **50**, 158-169 (2003).
4. S. Yang, J. N. Onuchic, and H. Levine, *J. Chem. Phys.* **125**, 054910 (2006).
5. J. Xing, H. Wang, and G. Oster, *Biophys. J.* **89**, 1551-1563 (2005).
6. L. M. Rice, E. A. Montabana, and D. A. Agard, *PNAS* **105**, 5378-5383 (2008).
7. E. Nogales and H.-W. Wang, *Curr. Op. Struct. Bio.* **16**, 221-229 (2006).
8. R. B. Ravelli et al, *Nature* **428** 198-202, (2004).
9. D. N. Drechsel and M. W. Kirschner, *Curr. Bio.* **4**, 1053-1061 (1994).
10. V. VanBuren, D. J. Odde, and L. Cassimeris, *PNAS* **99**, 6035-6040 (2002).
11. T. Schlick, E. Barth, and M. Mandziuk, *An. Rev. Bio. Biomol. Struc.* **26**, 181-222 (1997)
12. S. J. Marrink et al, *J. Phys. Chem. B*, **111**, 7812-7824 (2007).
13. B. M. Forrest and U. W. Suter, *J. Chem. Phys.* **102**, 7256-7266 (1995).

Mater. Res. Soc. Symp. Proc. Vol. 1274 © 2010 Materials Research Society　　　　1274-QQ06-03

Mechanics of collagen in the human bone: role of collagen-hydroxyapatite interactions

Shashindra M. Pradhan, Kalpana S. Katti and Dinesh R. Katti

Department of Civil Engineering, North Dakota State University, Fargo ND 58105 USA

ABSTRACT

Here, we report results of our simulations studies on modeling the collagen-hydroxyapatite (HAP) interface in bone and influence of these interactions on mechanical behavior of collagen through molecular dynamics and steered molecular dynamics (SMD). Models of hexagonal HAP ($10\bar{1}0$) and (0001) surface, and collagen with and without telopeptides were built to investigate the mechanical response of collagen in the proximity of mineral. The collagen molecule was pulled normal and parallel to the (0001) surface of hydroxyapatite. Water molecules were found have an important impact on deformation behavior of collagen in the proximity of HAP due to their large interaction energy with both collagen and HAP. Collagen appears stiffer at small displacement when pulled normal to HAP surface. At large displacement, collagen pulled parallel to HAP surface is stiffer. This difference in mechanical response of collagen pulled in parallel and perpendicular direction results from a difference in deformation mechanism of collagen. Further, the collagen molecule pulled in the proximity of HAP, parallel to surface, showed marked improvement in stiffness compared to absence of HAP. Furthermore, the deformation behavior of collagen not only depends on the presence or absence of HAP and direction of pulling, but also on the type of mineral surface in the proximity. The collagen pulled parallel to ($10\bar{1}0$) and (0001) surfaces showed characteristically different type of load-displacement response. In addition, here we also report simulations on 300 nm length of collagen molecule indicating the role of length of model on the observed response in terms of both the magnitude of modulus obtained as well as the mechanisms of response of collagen to loading.

INTRODUCTION

Collagen, the most abundant protein in the human body, is a structural protein that constitutes major portion of total bone protein. In human bone, the weight percentages of organic mineral phase, and water are about 25%, 65%, and 10% respectively [1-3]. The collagen molecule is a triple helical structure consisting of three poplypeptide chains (called α chains) wound into a right hand helix, and measures about 300 nm in length and 1.3 nm in diameter [4]. The polypeptide chains are made up of repeating triplets of (Glycine-X-Y), where X and Y are other amino acids. These three chains are held together by hydrogen bonds. In bone, the collagen molecules assemble in conjunction with mineral, water and other organics to form fibrils [5].

In this study, we have investigated directional dependence of collagen when pulled parallel and perpendicular to the mineral surface. The deformation behaviors of collagen in the proximity of different mineral surfaces are also studied. We have used (10 $\bar{1}$ 0) and (0001) surfaces of hydroxyapatite for our study. The detailed study of this can be found in Katti et al 2010[6]. We have also carried out steered molecular dynamics on a collagen of full length to investigate the role of length of molecule on elastic response and deformation mechanisms during loading.

MODEL BUILDING AND SIMULATIONS DETAILS

The model structure used for collagen consists of a GPP model of collagen (PDB ID: 1k6f) placed in the proximity of hydroxyapatite surface. The collagen structure is a triple helical molecule with telopeptides present at the ends. For our study, we have used collagen with N-terminal telopeptide as well as without telopeptide in the proximity of hydroxyapatite. The telopeptides are amino acid sequences present at the ends of collagen molecule. They are in the proximity of hydroxyapatite present in the hole zone. The structure of collagen with telopeptide was obtained from Malone et al[7].The sequence of N-terminal telopeptide used in our simulation is as follows[7-9]:

a1-GlnLeuSerTyrGlyTyrAspGluLysSerThrGlyIleSer-ValPro-helix (GlyProMet-)
a2-GlnPheAspAlaLysGlyGlyGlyPro-helix (GlyProMet-)

The structure of hydroxyapatite in bone is hexagonal with space group P6$_3$/m [10] and unit cell parameters a = 9.424 Å, b = 9.424 Å, c = 6.879, α = 90°, β = 90°, γ = 120°. The (0001) and (10 $\bar{1}$ 0) surfaces of hydroxyapatite were used in our models. These surfaces were created by repeating unit cells along appropriate axes. The detailed description of the methods used for preparing these surfaces are presented in our previous work [11-15].

The CHARMm (Brooks et al. 1983) force field is used for energy minimization and molecular dynamics. For solvation of collagen, TIP3P model of water is used. The temperature is increased gradually from 0 K to 300 K in steps of 100 K, and the pressure is increased in steps of 0.25 bar from 0 bar to 1.01 bar. Isothermal and isobaric ensemble was used for all the simulations. Particle Mesh Ewald (PME) was used for full electrostatics calculation and van der Waals cutoff distance of 10 Å was used for all the simulations. The steered molecular dynamics (SMD) is carried out on an equilibrated model by fixing one end of collagen and pulling its center of mass at constant velocity using a spring of force constant 4.0 kcal mol^{-1} Å^{-2}.

Besides the models described above, we also constructed a model collagen of full length. This long collagen molecule consists of 1014 residues along each chains is about 2900 Å in length. The collagen is solvated with 19417 water molecules. The collagen is minimized and equilibrated to 300 K and 1.01 bar using a method described previously. Steered molecular dynamics is carried out on the long collagen at different velocities using springs of force constants 0.1 and 1.0 kcal mol^{-1} Å^{-2}.

RESULTS AND DISCUSSION

The interaction between mineral and collagen is greatly influenced by the presence of water in the interface. The interaction energy evaluated between collagen, water, and hydroxyapatite show that water bears very high attractive interactions with both collagen and mineral (Fig 1). It is interesting to note that the interaction of water with collagen and mineral is much larger than the interaction between collagen and mineral. Because of this reason, water functions as an intermediary serving as a binding member between collagen and mineral as well as greatly influencing the deformation behavior of collagen.

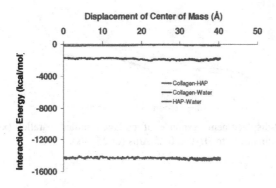

Figure 1: Interaction energies between collagen, HAP, and water when collagen is pulled

The load-displacement response of collagen pulled parallel and perpendicular to (0001) surface of HAP is shown is figure 2. The collagen pulled parallel to the HAP surface shows a stiffer response throughout the displacement range compared to the perpendicular direction. The collagen pulled perpendicular to HAP surface has to first overcome strong attraction of water molecules strongly bound to the HAP surface. This event dominates small displacement range of collagen pulled in the perpendicular direction resulting in relatively large initial stiffness compared to parallel direction of pulling. The hydrogen bonds that stabilize the triple helical structure of collagen continuously break as the collagen in stretched. The observation is that the initial breaking of hydrogen bonds is very rapid in case of collagen pulled parallel to the mineral surface as shown in figure 3. At slightly larger displacement the rate of breaking of hydrogen bonds in cases of pulling in parallel and perpendicular direction are comparable. With further pulling the unwinding of collagen becomes more prominent with the very rapid breaking of hydrogen bonds as shown in the figure. This rapid breaking stage is accompanied by increased stiffness of displacement response and increased interaction between water and collagen. The hydrogen bonding sites of collagen which are freed by breaking of inter-chain bonds interacts with water to form new hydrogen bonds in the vicinity. Additionally, the observation of load displacement response at pulling rates 0.25 Å/ps, 0.50 Å/ps, 0.75 Å/ps, and 1.0 Å/ps show the rate effects both in cases of collagen pulled in perpendicular as well as prarallel direction as shown in figure 4.

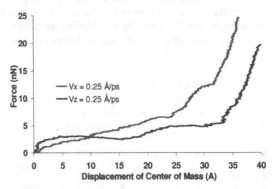

Figure 2: Load-displacement response of collagen pulled parallel (x-direction) and perpendicular (z-direction) to HAP at 0.25 Å/ps (or 25 m/s).

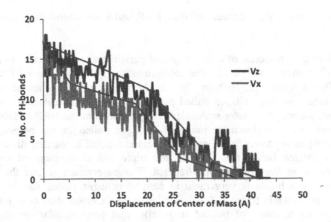

Figure 3: Evolution of hydrogen bonds when pulled along parallel (x-direction) and perpendicular (z-direction) to HAP

74

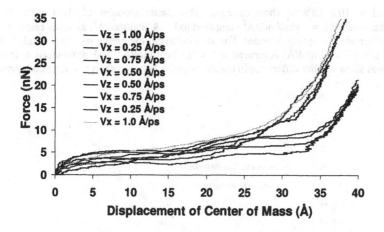

Figure 4: Load-displacement response of collagen pulled parallel (x-direction) and perpendicular (z-direction) to HAP at at different velocities.

The typical load displacement response of response of collagen pulled parallel to $(10\overline{1}0)$ and (0001) surfaces of HAP is shown in figure 5. At small displacement, higher stiffness is observed in case of collagen pulled in the proximity of (0001) surface compared to the $(10\overline{1}0)$ surface. At higher displacement, the stiffness responses of collagen pulled in the proximity of these surfaces are found to be comparable. Additionally, the load-displacement response of collagen pulled parallel to (0001) surface show step-like features which are less prominent in case of $(10\overline{1}0)$ surface. These observations point to the possibility that the deformational mechanism of collagen in the proximity of these surfaces could be quite different.

In case of simulations involving long and short collagen molecules, strain rates ranging from 0.0003/ps to 0.012/ps were used. The typical deformation response of short and long collagen molecules at this pulling rate is shown in figure 6. In this plot, the displacement of center of mass has been expressed as a percentage of initial length for easy comparison between short and long collagens; we refer to this ratio as reduced-displacement. In the initial region of load displacement curve, we observe straightening of a slightly coiled collagen molecule under applied tension. The elastic modulus is calculated by using a linear fit of the initial 7% of the load-displacement plot. The elastic modulus of short and long collagen molecules evaluated from SMD simulations at strain rate 0.012/ps are shown in table 1. For making a reasonable comparison between short and long collagen, we consider elastic modulus corresponding to $k = 4.0$ kcal/mol/Å^2 and $k = 0.1$ kcal/mol/Å^2 respectively for short and long collagen. It is seen that at the applied rate of pulling the elastic modulus for long collagen is 14.2 GPa, which is quite large compared to 10.8 GPa of short collagen. The elastic modules of short collagen is somewhat small at k = 1 kcal/mol/Å^2 compared to k = 4 kcal/mol/Å^2, as softer response is expected for smaller spring constant. For long collagen, the elastic modulus is relatively large at k = 1.0 kcal/mol/Å^2 compared to k = 0.1 kcal/mol/Å^2. The simulations at lower strain rates show smaller difference in elastic modulus between short and long collagens.

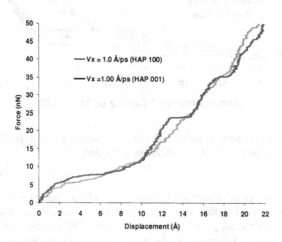

Figure 5: Force-displacement plot of collagen pulled paralled to (10$\overline{1}$0) and (0001) surfaces.

Table 1: Elastic modulus of short and long collagens molecules obtained for different values of spring constant.

k (kcal/mol/Å^2)	Short E (GPa)	Long E (GPa)
0.1	-	14.23
1	9.59	22.15
4	10.81	-

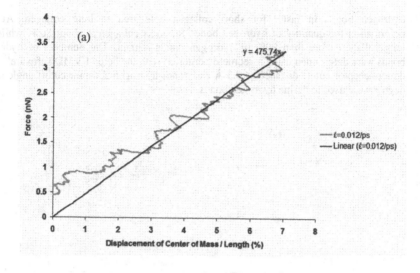

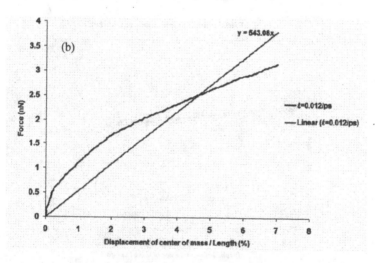

Figure 6: Load-displacement response of (a) short collagen and (b) long collagen.

The hydrogen bonds between polypeptide chains of collagen are the stabilizing force for the triple helical structure of collagen. As the collagen molecule is pulled, the numbers of hydrogen bonds between the chains fluctuate with deformation of collagen as seen in figure 7. For short collagen, the amplitude of fluctuation in number of hydrogen bonds is larger compared to long collagen. Additionally, the rate of depletion in number of hydrogen bonds in faster for short collagen compared to long collagen. At 7% deformation the quantity of hydrogen bonds for short collagen is about 80%, while for long collagen more than 90% of hydrogen bonds remain. The numbers of hydrogen bonds were determined using a geometric criterion with the help of VMD software[16]. The donor-acceptor cutoff distance of 3.5 Å and donor-hydrogen-acceptor cutoff angle of 60 degree were used to define hydrogen bonds.

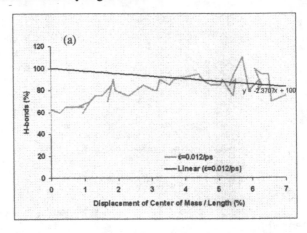

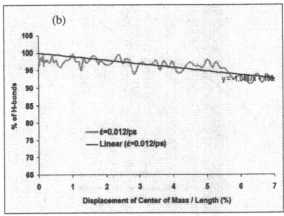

Figure 7: The evolution hydrogen bond during deformation for (a) short collagen and (b) long collagen.

CONCLUSIONS

The deformation mechanism of collagen is dependent on number of factors including strain rates, direction of pulling with respect to the mineral, and the type of mineral surface in the proximity. Additionally the length of collagen used in molecular dynamics models also has an important impact in observed in elastic response and deformation mechanism of collagen. Future work from our group addresses the role of length of collagen models on its mechanics, a factor fairly important for choice of range of velocities used. In this work, short collagen models have been used in proximity of hydroxyapatite which may have some observed differences from behavior of longer collagen models.

REFERENCES

1. J. E. Eastoe and B. Eastoe, *Biochem.* J. **57,** 453 (1954).

2. J. D. Termine, R. E. Wuthier, and A. S. Posner, *Proc Soc Exp Biol Med* **125,** 4 (1967).

3. J. D. Currey, P. Zioupos, P. Davies, and A. Casino, *Proc Biol Sci* **268,** 107 (2001).

4. W. J. Landis, M. J. Song, A. Leith, L. McEwen, and B. F. McEwen, *J Struct Biol* **110,** 39 (1993).

5. S. Uzel, M.J. Buehler, *Integrative Biology* **1,** 7 (2009).

6. D.R.Katti, S.M.Pradhan, and K.S.Katti, *Journal of Biomechanics* **43,** 9 (2010).

7. J. P. Malone, A. George, and A. Veis, *Proteins: Struct, Funct, Bloinf* **54,** 206 (2004).

8. E. Y. Jones and A. Miller, *Biopolymers* **26,** 463 (1987).

9. A. Otter, G. Kotovych, and P. G. Scott, *Biochemistry* **28,** 8003 (1989).

10. K. Sudarsanan and R. A. Young, *Acta Crystallogr.* , Sect. B **25,** 1534 (1969).

11. R. Bhowmik, K. S. Katti, and D. Katti, *Polymer* **48,** 664 (2007).

12. R. Bhowmik, K. S. Katti, and D. R. Katti, *J. Mater. Sci.* **42,** 8795 (2007).

13. R. Bhowmik, K. S. Katti, D. R. Katti, *International J. Nanotech.* **6,** 511 (2009).

14. R. Bhowmik, K.S. Katti, D. R. Katti, *J. Engineering Mechanics-ASCE,* **135,** 413 (2009).

15. R. Bhowmik, K.S. Katti, D. R. Katti, *J. Nanosci. Nanotech.* **8,** 2075 (2009).

16. W. Humphrey, A. Dalke, and K. Schulten, *J. Molec. Graphics,* **14.**1 (1996).

Mater. Res. Soc. Symp. Proc. Vol. 1274 © 2010 Materials Research Society   1274-QQ06-04

Humidity induced softening leads to apparent capillary effect in gecko adhesion

Bin Chen[*a], Huajian Gao[#]

[*]Engineering Mechanics, Institute of High Performance Computing, A*STAR,138632, Singapore.
[#]Division of Engineering, Brown University, Providence, RI, 02912, USA.

ABSTRACT

Humidity can lower the stiffness of beta keratin, which is the main component of spatula pads terminated at the bottom of a gecko's toe. To see if this softening can influence gecko adhesion, numerical simulation of the vertical peeling of a spatula pad adhered on a rough surface is performed, which shows that the reduction in material stiffness indeed leads to substantial increases in the pull-off force "(Chen, B. and Gao, H. International Journal of Applied Mechanics, 2010). This work provides an alternative explanation of experimental observations on the capillary effect in gecko adhesion.

[a]Corresponding author: Tel.: (65) 64191583; Email address: chenb@ihpc.a-star.edu.sg

INTRODUCTION

Among numerous conjectures on the underlying mechanism of gecko adhesion, such as glue, suction and interlocking, the van der Waals and capillary interactions have survived the experimental test so far. While it has been shown that the van der Waals force plays the primary role in gecko adhesion [2,3], humidity was also shown to have a strong effect [4-6].

Although the experiments have convincingly shown that humidity does play important roles in gecko adhesion, interpretation of the underlying physical mechanisms remains controversial. Fig. 1a shows the spatulae structure, the finest terminal branches of setae under gecko's toe. It is known that capillary forces due to liquid bridges would require a sufficient amount of water between gecko's spatula and substrate. Considering the fact that the spatula is hydrophobic, significant capillary condensation is not expected to occur until a relative humidity (RH) level exceeding 90% [5]. In fact, the ellipsometry data [5] confirmed that only an adsorbed monolayer of water is present. Meanwhile, another important experiment [7] showed that the adhesion energy remains constant and is independent of relative humidity up to 60-70% RH even for hydrophilic surfaces with contact angles less than 10 degree, which was consistent with other reports of the capillary effect on adhesion at nanoscale [8,9]. These experiments have casted substantial doubt on the existing interpretation of the observed humidity effect on gecko adhesion in terms of capillary condensation. Therefore, it remains a puzzle how relative humidity can cause a continuous rise in the pull-off force of a single spatula pad as observed in experiments [5].

This paper is a summary of our recent numerical simulations [1] which showed that the moisture induced stiffness reduction of spatula leads to substantially higher adhesion force with magnitude comparable to those observed in experiments. Based on these simulations, we are proposing an alternative interpretation of the effect of humidity on gecko adhesion, without the controversial assumption of adhesion energy continuously rising with relative humidity.

ANALYSIS

As shown in Fig. 1b, an elastic pad, representing a single spatula, adheres on a surface via a periodic array of attachment sites spaced at period l_p. The periodic attachments simulate imperfect contact between the spatula and the substrate due to the roughness of the surface and/or spatula pad as well as possible contamination. The patch size of each attachment site is assumed to be half of l_p. The thickness of the pad is chosen as $H = 10nm$, reflecting the thickness of a spatula pad which typically varies from 5 nm to 20 nm. The van der Waals adhesive energy between the pad and the substrate is assumed to be $\gamma = 0.04 N/m$.

(a)

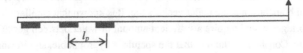

(b)

Figure 1 a) The spatulae are the finest terminal branches of setae on gecko's toe: SP, spatulae; BR,branch (adapted from [10]). b) An elastic spatula pad adhering on a rigid surface through periodic sites.

82

Abaqus/Standard is employed to simulate a spatula pad subjected to a vertical peeling force at one end as shown in Fig. 1b. The cohesive elements are used to model the adhesive interaction across the interface between the pad and substrate. The constitutive response of the cohesive elements is defined in terms of a traction-separation law. The damage of cohesive elements is initiated when the maximum of the nominal stress reaches 40 MPa. Damage evolution is based on an isotropic dependence of adhesion energy on the mode mix ratio. Two different values of the attachment period, $l_p = 10$ nm and $l_p = 16$ nm, are considered in the simulations. The stiffness of the pad under investigation varies from 0.4 GPa to 4GPa, which roughly corresponds to the observed range of stiffness changes of keratin due to moisture [11].

The simulated vertical peel-off force [1] increases substantially as the stiffness of the pad decreases. When $l_p = 10$ nm, the peel-off force increases from 0.023 N/m at a stiffness of 2 GPa to 0.04 N/m when the stiffness is reduced to 0.4 GPa. The change of stiffness between 2 GPa to 3 GPa does not affect the peel-off force significantly. When $l_p = 16$ nm, the peel-off force increases from 0.031 N/m at a stiffness of 4 GPa to 0.04 N/m when the stiffness is reduced to 0.75 GPa. When the stiffness is smaller than 0.75 GPa, the peel-off force seems to approach a saturation value of 0.04 N/m.

We point out that the above simulation results cannot be directly understood from Kendall's [12] model. According to Kendall's model, the vertical peel-off force of the pad can be expressed as

$$P_K = \frac{2\gamma}{\sqrt{1 + \frac{2\gamma}{EH}} + 1} \tag{1}$$

where γ is the adhesion energy of the interface. Since the stiffness of the pad under consideration is on the order of 1 GPa, the thickness H is on the order of 10 nm and the adhesion energy γ is on the order of 0.04 J/m^2, we see immediately that $\gamma / EH \ll 1$ and the pull-off force P_K is close to γ.

Our simulation results can only be understood based on our recent studies [13,14] of the apparent adhesion energy of an interface as a function of the fracture process zone size along the interface, where the local adhesion energy was allowed to vary periodically along the interface with a peak value and an average value within a period. As the interface is detached by propagation of a crack upon loading, the size of the fracture process zone near the crack tip is a material parameter which generally depends on the adhesion energy and adhesion strength of the interface, as well as the elastic stiffness of the surrounding media. It was shown [13, 14] that the apparent adhesion energy of the interface is equal to the average adhesion energy of the interface if the fracture process zone size is much larger than the period of cohesive interaction but becomes the peak value of the local adhesion energy when the opposite is true. In general, the apparent adhesion energy of the interface can vary

83

from the average value to the peak value of the local adhesion energy. For the present case of a thin pad adhering on a substrate with periodic attachments subjected to vertical peeling, the size of the fracture process zone is [13,14]

$$l_c \sim \left(\frac{EH^3\gamma}{\bar{\sigma}^2} \right)^{1/4} . \qquad (2)$$

In deriving Eq. (2), a thin elastic beam was considered to adhere on a rigid substrate [13]. Equation (2) was obtained by equating the stress intensity factor induced by the cohesive forces within the cohesive zone to that due to the far-filed applied load. This scaling law was confirmed by numerical simulation [13], and further generalized to cohesive zone sizes in thin film structures [14]. According to Eq. (2), lower material stiffness would decrease the fracture process zone size, which in turn would result in higher $\bar{\gamma}$, which is the apparent adhesion energy of the interface with defects; the same is true even for interfacial delamination in bulk solids [13,14]. This larger $\bar{\gamma}$ would then induce higher pull-off force of the pad, which is consistent with our numerical simulation.

DISCUSSION

In the simulations [1], the adhesion energy was assumed to remain constant and independent of humidity but the material stiffness of β keratin is reduced by the presence of moisture, in a range according to the experimental observations [11]. The simulations showed that the lower stiffness results in substantially higher vertical pull-off force comparable to the experimental data. This analysis provides an alternative explanation of the observed humidity effect on gecko adhesion. We suggest that humidity does not directly affect the adhesion energy of the interface but rather change the apparent adhesion energy by reducing the stiffness of the spatula pad through moisture adsorption [1].

This explanation would not violate the experimental observation [7] that the adhesion energy between micromechined surfaces is not affected by relative humidity until 60-70% RH even for hydrophilic surfaces. On the other hand, Seta comprises of hard keratinous fibrils embedded in a soft matrix and the fibrils are mainly composed of β sheets [15], which can be strongly affected by humidity. It was shown [11] that the stiffness of ostrich feather rachis decreases from 3.66 GPa at 0% RH to 2.58 GPa at 50% RH and to 1.47 GPa at 100% RH, and the stiffness of an ostrich claw decreases from 2.7 GPa at 0% RH to 2.07 GPa at 50% RH and then to 0.14 GPa at 100% RH. With this range of relative humidity induced stiffness reduction in mind, the simulation results indicate that the pull-off force increases substantially with rising relative humidity, as observed in the experiments [5]. According to Eq. (2), the fracture process zone size would increase when the cohesive strength is reduced. A lower cohesive strength would make the apparent adhesion energy less sensitive to the roughness of the substrate, which also seems to agree with the observation [5] that the vertical pull-off force on more weakly interacting substrates showed smaller increases with relative humidity and larger scatters than that on glass substrate. We note that so far there has been no direct observation on the effects of relative humidity on the stiffness of seta. This is probably due to a thin protection layer of sheath surrounding seta [15]. This sheath layer is no longer present in the distal region of seta, which caused material in this region vulnerable to laser induced heating treatment [15]. Without this protection sheath, it seems indeed possible that the stiffness of spatulae can be significantly affected by humidity.

CONCLUSIONS

The present results show that the moisture induced stiffness reduction can result in substantially higher pull-off force [1]. Our analysis provides so far the only explanation of humidity enhanced gecko adhesion without involving the capillary effect [1]. This work suggests a distinct possibility that the van der Waals force could be the only mechanism for gecko adhesion, and that the apparent humidity effect could be a sophisticated manifestation of van der Waals interaction via stiffness changes in spatula in the presence of humidity [1]. In the future, it will be interesting to investigate the humidity effect in adhesion by molecular dynamics simulations, similar to previous studies on the size effect [16].

REFERENCES

1. Chen, B. and Gao, H., 2010 An alternative explanation of the effect of humidity in gecko adhesion: stiffness reduction enhances adhesion on a rough surface. *International Journal of Applied Mechanics. 2, 1-9.*
2. Autumn, K., Liang, Y.A., Hsieh, S.T., Zesch, W., Chan, W.P., Kenny, T.W., Fearing, R., and Full, R.J., 2000 Adhesive force of a single gecko foot-hair *Nature* 405, 681-685.
3. Autumn, K., Sitti, M., Liang, Y.A., Peattie, A.M., Hansen, W.R., Sponberg, S., Kenny, T.W., Fearing, R., Israelachvili, J.N., and Full, R.J., 2002 Evidence for van der Waals adhesion in gecko seta. *Proc. Natl. Acad. Sci. USA* 99, 12252-12256.
4. Sun, W., Neuzil, P., Kustandi, T.S., Oh, S., and Samper, V.D., 2005 The nature of the gecko lizard adhesive force. *Biophy. J.* 89, L14-L17.
5. Huber, G., Mantz, H., Spolenak, R., Mecke, K., Jacobs, K., Gorb, S.N., and Arzt, E., 2005 Evidence for capillarity contributions to gecko adhesion from single spatula nanomechanical measurements. *Proc. Natl. Acad. Sci. USA.* 102, 16293-16296.
6. Niewiarowski, R. H., Lopez, S., Ge, L., Hagan, E., Dhinojwala, A., 2008 Sticky Gecko feet: the role of temperature and humidity. *PLoS ONE*, 3, 2192.

7. Delrio, F. W., Dunn, M. L., Phinney, L. M., Bourdon, C. J., De boer, M. P.,2007 Rough Surface adhesion in the presence of capillary condensation. *Applied Physics Letters*, 90,163104.

8. Xiao, X., Qian, L., 2000 Investigation of humidity-dependent capillary force. *Langmuir* 16: 8153-8158.

9. He, M., Blum, A.S., Aston, D.E., Buenviaje, C., Overney, R.M., Luginbuhl, R. J., 2001 Critical phenomena of water bridges in nanoasperity contacts. *J. Chem. Phys.* 114, 1355-1360.

10. Gao, H., Wang, X., Yao, H., Gorb, S., and Arzt, E., 2005 Mechanics of hierarchical adhesion structures of geckos. *Mech. Mater.* 37. 275–285.

11. Taylor, A. M., Bonser, R.H.C., Farrent, J. W., 2004 Influence of hydration on the tensile and compressive properties of avian keratinous tissues. *Journal of Materials Science* 39, 939.

12. Kendall, K., 1975 Thin-film peeling-elastic term. *J. Phys. D: Appl. Phys.* 8, 1449-1452.

13. Chen, B., Shi, X.H., and Gao, H., 2008 Apparent fracture/adhesion energy of interfaces with periodic cohesive interactions. *Proc. R. Soc. A.* 464, 657-671.

14. Chen, B., Wu, P.D., and Gao, H., 2009 Geometry- and velocity- constrained cohesive zones and mixed mode fracture/adhesion energy with periodic cohesive interactions. *Proc. Roy. Soc. A* 465, 1043-1053.

15. Rizzo, N.W., Gardner, K.H., Walls, D.J., Keiper-Hrynko, N.M., Ganzke, T.S., and Hallahan, D.L., 2006 Characterization of the structure and composition of gecko adhesive setae. *J. R. Soc. Interface* 3(8), 441-451.

16. Buehler, M.J., Yao, H., Gao, H., and Ji, B., 2006 Cracking and adhesion at small scales: atomistic and continuum studies of flaw tolerant nanostructures *Modelling Simul. Mater. Sci. Eng.* 14, 799.

Structural Materials and Tissues I

Mater. Res. Soc. Symp. Proc. Vol. 1274 © 2010 Materials Research Society   1274-QQ07-03

Computational multiscale studies of collagen tissues in the context of brittle bone disease *osteogenesis imperfecta*

Alfonso Gautieri[1, 2], Simone Vesentini[2], Alberto Redaelli[2] and Markus J. Buehler[1,]*

[1]Laboratory for Atomistic and Molecular Mechanics, Department of Civil and Environmental Engineering, Massachusetts Institute of Technology, 77 Massachusetts Ave. Room 1-235A&B, Cambridge, MA, USA.

[2]Biomechanics Group, Department of Bioengineering, Politecnico di Milano, Via Golgi 39, 20133 Milan, Italy.

* Corresponding author, mbuehler@MIT.EDU

ABSTRACT

Osteogenesis imperfecta (abbreviated as OI) is a genetic disorder in collagen characterized by mechanically weakened tendon, fragile bones, skeletal deformities and in severe cases prenatal death. Even though many studies have attempted to associate specific mutation types with phenotypic severity, the molecular and mesoscale mechanisms by which a single point mutation influences the mechanical behavior of tissues at multiple length-scales remain unknown. Here we review results of a hierarchy of full atomistic and mesoscale simulations that demonstrated that OI mutations severely compromise the mechanical properties of collagenous tissues at multiple scales, from single molecules to collagen fibrils. Notably, mutations that lead to the most severe OI phenotype correlate with the strongest effects, leading to weakened intermolecular adhesion, increased intermolecular spacing, reduced stiffness, as well as a reduced failure strength of collagen fibrils (Gautieri *et al.*, *Biophys. J.*, 2009). Our study explains how single point mutations can control the breakdown of tissue at much larger length-scales, a question of great relevance for a broad class of genetic diseases. Furthermore, by extending the MARTINI coarse-grained force field, we provide a new modeling tool to study collagen molecules and fibrils at much larger scales than accessible to existing full atomistic models, while incorporating key chemical and mechanical features and thereby presents a powerful approach to computational materiomics (Gautieri *et al.*, *Journal of Chemical Theory and Computation*, 2010). We describe the coarse-graining approach and present preliminary findings based on this model in applying it to large-scale models of molecular assemblies into fibrils.

INTRODUCTION

Collagen is a crucial structural protein material, formed through a hierarchical assembly of tropocollagen molecules, arranged in collagen fibrils that constitute the basis for larger-scale fibrils and fibers. The contour length of collagen molecules is ≈300 nm, which represents a scale that is not yet accessible to full atomistic molecular modeling. Furthermore, the study of collagen fibrils and fibers is currently not feasible with full atomistic simulation, which represents an important limitation since this scale in the structural hierarchy of collagen is most relevant for physiological function as well as for the onset of diseases like the so-called "brittle bone disease" or *osteogenesis imperfecta* (OI), a disease due to glycine mutations which lead to a catastrophic breakdown of collagen based tissues [1]. Earlier experimental studies have attempted to correlate glycine mutation types and locations with phenotypic severity. Some trends are apparent, such as OI severity increasing with an amino to carboxyl terminal orientation and with substitution by large and charged amino acids. At present, however, genotype-phenotype correlations are too weak to accurately predict the phenotypic effect of a particular glycine mutation (Figure 1a). The answer to the question how a single point mutation in the tropocollagen molecule can cause failure of collagen based protein materials remains unknown. In particular, it remains unclear at what level in the tissue structure and how a single point mutation influences the behavior. To address these points, investigations at all relevant hierarchical levels must be carried out, beginning at the molecular level. Here we review a series of systematic molecular scale bottom-up computational experiments focused on pure collagenous tissue, carried out using atomistic-level Molecular Dynamics (MD), Adaptive Poisson-Boltzmann Solver (APBS) calculations, a mesoscale model and a coarse-grained approach for fibril modeling based on the MARTINI force field, which allows the study of complex molecular systems up to micrometer dimension and millisecond duration, while incorporating key biochemical features [9, 10, 12].

THEORY

Triple helical tropocollagen molecule models with various sequences are created using the THeBuScr (Triple-Helical collagen Building Script) tool [2]. All simulations are carried out with explicit solvent and the GROMOS96 43a1 force field as reported in earlier studies [3, 4].

Molecular scale modeling

We choose the simplest model of tropocollagen, with only Gly-Pro-Hyp (GPO) triplets on each of the three chains as the reference system (where Hyp and O are the three letter code and single letter code for the amino acid hydroxy proline, respectively). The central triplet is used to introduce the Gly replacement. The peptide structure is $[(GPO)_5-(XPO)-(GPO)_4]_3$, where the X position is occupied by alanine, serine, cysteine, arginine, valine, glutamic acid or aspartic acid (Figure 1b). The mechanical properties (Young's modulus) of each collagen peptide are determined through Steered Molecular Dynamics simulations [3]. The electrostatic interactions between the equilibrated peptides are calculated through APBS calculations. The APBS calculations involve, for each pair of peptides (p) and for each relative distance (r), the evaluation of the electrostatic energy of three systems: both single peptides (C_a and C_b) and the complex (C_c); the electrostatic interaction energy (C_{int}) is then calculated as: $C_{int}(p,r)=C_c-(C_a+C_b)$. We calculate the van der Waals energy following a procedure similar to that described for the electrostatic interaction calculations. The total interaction energy is then obtained as the sum of the electrostatic interaction energy and the van der Waals components. These profiles are then fitted to a 12:6 Lennard-Jones potential and the values of ε (energy well depth) and d_{min} (distance between the particles at the energy minimum) are extracted for each peptide pair.

Fibril scale modeling

The mesoscale model of collagen fibrils is identical to that reported in [5, 6], with minor modifications to reflect changes in the segment's stiffness and intermolecular interactions due to OI mutations. The mesoscale model is based on the concept to treat a tropocollagen molecule as a linear string of beads, where individual beads interact according to a stretching and bending potential that reflect the physical properties of tropocollagen molecules. Finally, in order to overcome the limitation of the mesoscale model (simplified representation of structure and sequence), we use a computational multiscale approach to generate an atomistic-informed coarse-grained model of tropocollagen in the framework of the MARTINI force field. The parameters of the coarse-grained model are obtained through a combination of experimental and full-atomistic modeling data to identify the parameters for hydroxyproline residues, as well through a statistical analysis of collagen-like PDB entries to assess the geometrical features of the coarse-grained tropocollagen triple helix. The results of the parameterization are fed into the coarse-grained collagen model in the spirit of a multi-scale simulation approach (Figure 2a).

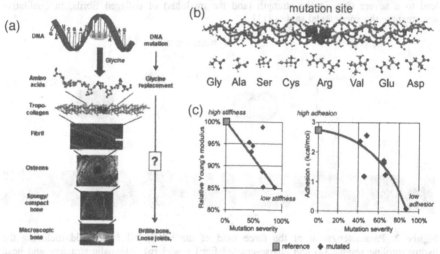

Figure 1. Overview over different material scales in bones (a), showing hierarchical collagen structure from collagen molecule to macroscopic level (left) and known effects of OI at each scale (right). Tropocollagen atomistic model (b), indicating the residue that is replaced in the mutation (thick red). OI mutations lead to a slightly lower mechanical properties at single molecule scale (panel (c), left) and moderate to severe decrease in the adhesion energy at the intermolecular scale (panel (c), right).

RESULTS AND DISCUSSION

Molecular scale effects

The Young's modulus of a single reference tropocollagen molecule is determined to be 3.96 GPa, in agreement with corresponding experimental results [7, 8]. Molecules with glycine mutations display slightly softer mechanical properties than the reference peptide, with values ranging from 3.37 GPa to 3.78 GPa. A decrease in Young's modulus up to 15% is observed, depending on the specific type of the replacing residue (left in Figure 1(c), see also [9]). We proceed with a systematic investigation of intermolecular interactions, for all seven mutations as well as the reference system, by studying the adhesion energy profile. We find that the intermolecular adhesion is strongly influenced by the type of mutation in the interacting tropocollagen molecules [10]. When the reference tropocollagen molecule is coupled to an OI

91

mutated peptide, the adhesion energies are reduced compared to the case when the reference peptide is interacting with another reference peptide without OI mutation. In addition, the energy minimum is shifted to larger distances, leading to an increased intermolecular spacing. A quantitative analysis of OI effects on intermolecular adhesion is carried out by plotting the depth of the energy minimum ε (equivalent to the adhesion energy) as a function of the severity of each mutation type (Figure 1c, right). The adhesion energy ε decreases from 2.8 kcal/mol (reference case) to 0.1 kcal/mol (most severe mutation), reflecting a significant reduction in intermolecular adhesion by more than 90%. In our studies we have demonstrated that these local changes can lead to a severe drop in the strength (and the modulus) of collagen fibrils, in qualitative agreement with experimental studies [11].

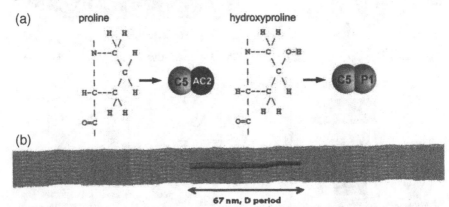

Figure 2. Parameterization of the force field of the MARTINI force field including the hydroxyproline residue (a) and coarse-grained fibril model (b). Atomistic structure and bead representation using MARTINI model notation of proline are shown on left side of panel (a), while atomistic structure and MARTINI bead representation of hydroxyproline are shown on the right side of panel (a), while the bead representation using MARTINI model notation are shown on the right side. The fibril model (b), when showing the period images of the unit cell (at the center, periodic images in blue), reproduce the typical banding seen in AFM images and due to the staggering of collagen molecules.

Fibril scale effects

The mesoscale model provides us with the ability to directly study the properties of collagen fibrils (at scales of larger than micrometers) due to molecular-level, single point mutations (at scales of sub-nanometers). We find that the presence of mutations leads to a significant change in the mechanical response of collagen fibrils [10]. This level of coarse-graining, however, is at a relatively coarse level where information about biochemical features (e.g. amino acid sequence) cannot be represented directly. However, the incorporation of biochemical features is crucial in a computational materiomics approach where material properties are elucidated at multiple scales.

The purpose of the coarse-grain approach is here to extend collagen modeling to larger structures (e.g., the collagen fibril) and larger time intervals without the simplifications of the mesoscale model described above but with the computational advantage of coarse-grain approach (reduced particles number and larger time step, leading to an overall speedup of up to

92

400). Several validation computations have been carried out to ensure that the extended MARTINI model can accurately describe key mechanical and biophysical parameters of collagen molecules (Figure 2a) [12]. First, we considered the Young's modulus of a collagen molecule by simulating a short collagen-like peptide. The collagen molecule was subjected to axial load by using steered molecular dynamics, and the study resulted in a Young's modulus value of 4.62 GPa, which is in good agreement with those obtained both through atomistic setups [5, 13-15] and experimental analysis [16]. Second, we have computed the persistence length of collagen molecules, leading to L_p = 51.5±.6.7 nm. We find reasonable agreement with results from earlier studies. Due to its contour length of 300 nm, all-atom simulations with explicit solvent are prohibitive since they would require excessive computational resources due to the very large number of particles. The reduction by roughly a factor of 10 in the coarse-grained description provides a significant speedup that facilitates the direct simulation of much longer molecules. Furthermore, the typical time step used in classical molecular dynamics simulations (1-2 fs) allows the modeling only on the nanoseconds time scale. Whereas the coarse-grained model enables one to use much longer time-scales on the order of 20-40 fs. Considering the combined effect of the reduction of the number of particles and the increased time step, the coarse-grained approach leads to a total speed-up of 200-400 with respect to atomistic simulations.

CONCLUSIONS

With the coarse-grain significant computational speedup, the modeling framework reported here opens many possibilities for future studies, particularly at the scale of collagen fibrils (Figure 2b) and possibly fibers. The coarse-grain approach will facilitate the computational investigation of how changes in the sequence would influence the packing of collagen fibril, helping the understanding of the mechanisms underlying collagen-related diseases, such as *osteogenesis imperfecta*. This, together with the study of the interaction with other relevant biomolecules such as proteoglycans, will provide useful details for the understanding of structure-properties relationship in the broader class of collagen materials. The study of materials phenomena in the context of injury and disease is a powerful approach that could have implications for the development of novel treatment options through innovative methods to intervene with disease mechanisms at a fundamental level [17]. Future work could include the consideration of other amino acid sequences (and compositions, to investigate physiological and disease states), and the addition of mineral crystals to provide a molecular-level model of mineralized collagen or bone.

ACKNOWLEDGMENTS

This research was supported by the National Science Foundation, and in part through the MIT-Italy program (Progetto Rocca).

REFERENCES

1.   Peltonen, L., A. Palotie, and D.J. Prockop, *A defect in the structure of type I procollagen in a patient who had osteogenesis imperfecta: excess mannose in the COOH-terminal*

propeptide. Proceedings Of The National Academy Of Sciences Of The United States Of America, 1980. **77**: p. 6179-6183.

2. Rainey, J. and M. Goh, *An interactive triple-helical collagen builder.* Bioinformatics, 2004. **20**(15): p. 2458-2459.

3. Gautieri, A., M.J. Buehler, and A. Redaelli, *Deformation rate controls elasticity and unfolding pathway of single tropocollagen molecules.* Journal of the Mechanical Behavior of Biomedical Materials, 2009. **2**(2): p. 130-137.

4. Gautieri, A., et al., *Mechanical properties of physiological and pathological models of collagen peptides investigated via steered molecular dynamics simulations.* Journal Of Biomechanics, 2008. **41**(14): p. 3073-3077.

5. Buehler, M.J., *Nature designs tough collagen: Explaining the nanostructure of collagen fibrils.* P. Natl. Acad. Sci. USA, 2006. **103**(33): p. 12285–12290.

6. Buehler, M.J., *Nanomechanics of collagen fibrils under varying cross-link densities: Atomistic and continuum studies.* Journal of the Mechanical Behavior of Biomedical Materials, 2008. **1**(1): p. 59-67.

7. Sun, Y.L., et al., *Stretching type II collagen with optical tweezers.* Journal Of Biomechanics, 2004. **37**(11): p. 1665-1669.

8. Sasaki, N. and S. Odajima, *Stress-strain curve and Young's modulus of a collagen molecule as determined by the X-ray diffraction technique.* Journal of Biomechanics, 1996. **29**(5): p. 655-658.

9. Gautieri, A., et al., *Single molecule effects of osteogenesis imperfecta mutations in tropocollagen protein domains.* Protein Science, 2009. **18**(1): p. 161-168.

10. Gautieri, A., et al., *Molecular and Mesoscale Mechanisms of Osteogenesis Imperfecta Disease in Collagen Fibrils.* Biophysical Journal, 2009. **97**(3): p. 857-865.

11. Misof, K., et al., *Collagen from the osteogenesis imperfecta mouse model (oim) shows reduced resistance against tensile stress.* Journal Of Clinical Investigation, 1997. **100**(1): p. 40-45.

12. Gautieri, A., et al., *Coarse-Grained Model of Collagen Molecules Using an Extended MARTINI Force Field.* Journal of Chemical Theory and Computation, 2010. **6**(4): p. 1210-1218.

13. Vesentini, S., et al., *Molecular assessment of the elastic properties of collagen-like homotrimer sequences.* Biomechanics And Modeling In Mechanobiology, 2005. **3**(4): p. 224-234.

14. Buehler, M.J., *Atomistic modeling of elasticity, plasticity and fracture of protein crystals.* Journal of Computational and Theoretical Nanoscience, 2006. **3**(5): p. 670–683

15. Lorenzo, A.C. and E.R. Caffarena, *Elastic properties, Young's modulus determination and structural stability of the tropocollagen molecule: a computational study by steered molecular dynamics.* Journal Of Biomechanics, 2005. **38**(7): p. 1527-1533.

16. Sasaki, N. and S. Odajima, *Elongation mechanism of collagen fibrils and force-strain relations of tendon at each level of structural hierarchy.* Journal Of Biomechanics, 1996. **29**(9): p. 1131-1136.

17. Buehler, M.J. and Y.C. Yung, *Deformation and failure of protein materials in physiologically extreme conditions and disease.* Nat Mater, 2009. **8**(3): p. 175-88.

Mater. Res. Soc. Symp. Proc. Vol. 1274 © 2010 Materials Research Society   1274-QQ07-04

Effect of environment on mechanical properties of micron sized beams of bone fabricated using FIB

Ines Jimenez-Palomar and Asa H. Barber

Department of Materials, School of Engineering and Materials Science
Queen Mary University of London
London E1 4NS
United Kingdom.

ABSTRACT

Bone is a complex material with structural features varying over many different length scales. Lamellae in bone are discrete units of collagen fibril arrays that are the dominant structural feature at length scales of a few microns. The mechanical properties of bone are importantly dependent on the synergy between the lamellae and structural features at other length scales. However, the mechanical properties at this micron level will be indicative of the bone material itself and ignores the structural and geometric organizations prevalent at larger length scales. The isolation of volumes of bone at the lamellar level requires precision cutting methodology and this paper exploits Focused Ion Beam (FIB) methods to mill small cantilever beams from bulk bone material. Importantly, FIB milling can only be performed in a relatively high vacuum environment. Atomic Force Microscopy (AFM) mechanical tests are therefore performed in two environments, high vacuum and air in order to assess the effects of vacuum on bone beam mechanical behaviour. Our results indicate that little difference in the bone beam elastic modulus is found from bending experiments at deflections up to 100nm in different environments.

INTRODUCTION

Bone is a complex material with structural features varying over many different length scales. The lamellar unit in bone is a common feature at the micron level and can be considered as the building block of cortical bone.[1] This lamellar unit has been shown to be composed of five subunits, with each subunit composed of an array of aligned mineralized collagen fibrils. The orientation of the sublayers has been shown to conform to a rotated plywood-like structure.[2] A number of researchers[1-3] have shown that the lamellar unit can be divided into two subunits; the 'thick' subunit where the collagen fibrils run parallel or at 30° to the long axis, thus contributing significantly to the elastic modulus,[4] and the 'thin' subunit for fibrillar arrays oriented at 60° and 90° to the long axis. Indeed, the collagen and lamellar organization has been previously identified as significant for defining mechanical properties of bone, such as the anisotropy found in human cortical bone [3] and lamellar bone,[2] although the relationship between mechanical performance at the lamellar micron to sub-micron length scales and how this related to overall bone mechanical behaviour is unclear.

The mechanical properties of individual lamellae have been previously measured using a range of different methods. Ascenzi[4] isolated individual lamellae by detaching whole circular lamellar rings followed by unrolling in order to perform direct tensile testing on the individual sheets. However, the unrolling of lamellar rings is potentially evasive and will cause bending stresses within the material that may initiate cracking, although Ascenzi indicated that the stresses were not sufficient to cause cracking.[4] Further mechanical investigations of bone have been carried out by Hasegawa,[5] Turner[6] and Pidaparti[7] using acoustic microscopy. However, the measured mechanical properties of the bone material were recorded over many tens of microns and therefore evaluated a number of lamellae. Nanoindentation remains perhaps the most suitable to measure the mechanical properties of bone within a relatively small, and defined, volume.[1,8] However, nanoindentation causes a complex stress condition within the bone that may not be indicative of the stress conditions that the bone structure is adapted towards. This paper uses a novel method of FIB to isolate discrete volumes of bone and subsequently mechanically test using AFM. As the bone volumes examined are relatively small, AFM has sufficient force resolution to potential elucidate complex mechanical deformation processes. Thus, testing of bone at micron length scales can provide direct information on the mechanical properties of the bone material, especially as the intrinsic mechanical properties of bone are dependent on the lamellar structure.[2] In addition, the high vacuum conditions required for FIB milling and its effects on the mechanical properties of bone at the micron level are systematically investigated by comparing to other environmental conditions.

EXPERIMENTAL

The mechanical properties of individual micron-sized units of bone were evaluated by isolating individual beams from the parent material using a multi-stage process. Frozen Equine metacarpals were thawed and cut using a circular saw into 2x2x2mm cubes as shown in Figure 1. Each cube was embedded in polymer resin (EpoFix Kit, Struers, Denmark), cured and polished using increasing Emery paper grades (3M, USA) and diamond paste. This polishing allowed osteonal organization to be observed and identification of lamellar rings using polarized light microscopy and scanning electron microscopy (SEM) equipped with a Back Scatter Electron Detector (BSED). After imaging, the sample was cut out of the resin and dehydrated using increasing concentrations of alcohol in water in order to prepare them for FIB milling in vacuum. The sample needed to be dehydrated in order to avoid charging effects that could interfere with the milling process and to prevent drying cracks that could be caused by the vacuum.

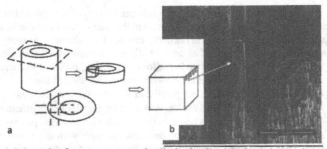

Figure 1 – a) Schematic of transverse cut at the diaphysis, sliced further using a microtome (left). Final sample with the beams milled using FIB (middle). b) SEM image of an individual bone beam FIB milled from the bulk bone with the beam's long axis parallel to the long axis of bone.

96

FIB milling of small beams (2x2x10μm) was carried out using a dual-beam system (Quanta 3D, FEI, EU/USA) combining the FIB with SEM imaging. FIB milling at 30kV and 1nA was carried out in the region of the lamellar rings in order to fabricate an individual beam from the lamellar region as shown in Figure 1(b). The FIB milled beams were therefore representative of an individual lamella. Mechanical testing of the individual bone beams shown in Figure 1(b) were performed using AFM in two different configurations: i) beam bending at high vacuum (5.25x10^{-4} Pa) using a custom build AFM (Attocube GmbH, Ger.) situated in the vacuum chamber of the dual-beam system ii) removal of the sample from the vacuum chamber and beam bending in air using a commercially available AFM (NTegra, NT-MDT, Rus.).

RESULTS AND DISCUSSION

Mechanical testing was performed by contacting the end of the bone beam with the AFM probe as shown in the schematic in Figure 2. The force applied by the AFM probe causes the bone beam to bend by a measurable deflection, with the applied force and beam deflection recorded using the AFM. The elastic modulus of a deflected cantilever beam can be found using Euler–Bernoulli beam theory by measuring the dimensions of the beam and the applied load P as shown in Figure 2.

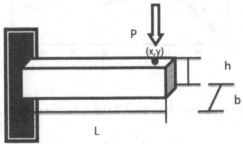

Figure 2 - Schematic of a rectangular beam of Length L, breadth b and height h bending under an applied load P.

For a bone beam of length L, breadth b and thickness h, the elastic modulus of the bone beam E can be related to the recorded beam deflection δ and applied load P by:

$$E = \frac{12L^3}{3bh^3} \cdot \frac{P}{\delta}$$

Equation 1

The above equation can be rearranged to show how the applied stress in the beam σ varies with deflection δ by:

97

$$E = \frac{2L^2}{3h} \cdot \frac{\sigma}{\delta}$$ Equation 2

For a bone beam of the same dimensions, the stress in the beam verses deflection δ is shown in Figure 3 below.

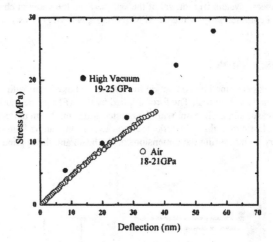

Figure 3 - Plot of the stress in the bone beam during deflection measuring from AFM bending experiments.

The initial linear region shown in Figure 3 at relatively small deflections up to 100nm is used to calculate the elastic modulus of the sample using Equation 2. Note that we refer to this initial region as elastic due to its linear behaviour although the bone may effectively be viscoelastic to some extent. The calculated elastic modulus from the fitting in Figure 3 suggests that there is little difference between the elastic modulus values for all of the beams although perhaps bending in high vacuum gives a slightly higher elastic modulus. This increase in elastic modulus with dehydration has been observed previously for large bone volumes of many cubic millimetres.[6] Thermal Gravimetric Analysis (TGA) was also used to corroborate the drying effect of the vacuum chamber. The weight gain between bone left in air and bone from the vacuum chamber was no more than 3 wt%, leading to the conclusion that the effect of the vacuum removes less water at this hierarchical level than studies examining larger bone volumes.[6] We can therefore state that, at relatively small beam bending deflections, there is little effect of the environment of the mechanical properties of the bone beams. We also note that the elastic modulus values measured from our beam bending experiments are similar to elastic modulus values from 3-point bending,[9] 4-point bending[10] and nanoindentation[11] of equine bone.

CONCLUSIONS

AFM bending tests were used to measure the elastic modulus of bone beams isolated from the parent cortical bone using FIB. The elastic moduli of these bone cantilevers showed little difference between the different environments and were similar to literature values. The results suggested that our preparation methods are sufficient to test bone in a quasi-native state. These preparation methods show potential of using a combination of AFM and SEM setup, as has been previously performed by our group,[12, 13] for sub lamellar cortical bone to be tested in a range of different mechanical testing modes such as in tension, buckling and shear.

REFERENCES

1. H. S. Gupta, U. Stachewicz and W. Wagermaier, J. Mater. Res. 21 (8), 1913-1921 (2006).
2. S. Weiner, W. Traub and H. D. Wagner, Journal of Structural Biology 126 (3), 241-255 (1999).
3. S. K. Boyd and B. M. Nigg, in *Biomechanics of the Musculo-skeletal System*, edited by B. M. Nigg and W. Herzog (Wiley, Canada, 2007), pp. 65-94.
4. A. Ascenzi, A. Benvenuti and E. Bonucci, Journal of Biomechanics 15 (1), 29-37 (1982).
5. K. Hasegawa, C. H. Turner and D. B. Burr, Calcified Tissue International 55, 381-386 (1994).
6. C. H. Turner and D. B. Burr, Calcified Tissue International 60, 90 (1997).
7. R. M. Pidaparti, A. Chandran, Y. Takano and C. H. Turner, Journal of Biomechanics 29 (909-916) (1996).
8. J. Y. Rho, T. Y. Tsui and G. M. Pharr, Biomaterials 18, 1325-1330 (1997).
9. G. Reilly and J. D. Currey, The Journal of Experimental Biology 202, 543-552 (1999).
10. G. Bigot, A. Bouzidi, C. Rumelhart and W. Martin-Rosset, Med. Eng. Phys. 18 (1), 79-87 (1996).
11. J. Y. Rho, J. D. Currey, P. Zioupos and G. Pharr, The Journal of Experimental Biology 204, 1775-1781 (2001).
12. F. Hang, D. Lu, S. W. Li and A. H. Barber, Probing Mechanics at Nanoscale Dimensions 1185, 87-91 (2009).
13. F. Hang, D. Lu and A. H. Barber, Structure-Property Relationships in Biomineralized and Biomimetic Composites 1187, 135-140 (2009).

Structural Materials and Tissues II

Mater. Res. Soc. Symp. Proc. Vol. 1274 © 2010 Materials Research Society   1274-QQ08-01

The Scissors Model of Microcrack Detection in Bone: Work in Progress

David Taylor[1], Lauren Mulcahy[1,2], Gerardo Presbitero[1], Pietro Tisbo[1], Clodagh Dooley[1], Garry Duffy[2] and T.Clive Lee[2]
[1]Trinity Centre for Bioengineering, Trinity College, Dublin 2, Ireland.
[2]Department of Anatomy, Royal College of Surgeons in Ireland, Dublin, Ireland. ·

ABSTRACT

We have proposed a new model for microcrack detection by osteocytes in bone. According to this model, cell signalling is initiated by the cutting of cellular processes which span the crack. We show that shear displacements of the crack faces are needed to rupture these processes, in an action similar to that of a pair of scissors. Current work involves a combination of cell biology experiments, theoretical and experimental fracture mechanics and system modelling using control theory approaches. The approach will be useful for understanding effects of extreme loading, aging, disease states and drug treatments on bone damage and repair; the present paper presents recent results from experiments and simulations as part of current, ongoing research.

INTRODUCTION

The idea for the so-called "scissors model" first came about in 2001: the concept has now appeared in a number of publications [1-5] and currently is being actively researched by a team which includes materials scientists, cell biologists, microscopists and experts in engineering control theory. In this paper we take the opportunity to present for the first time some of our most recent results, from experimental activities which are still ongoing. Because this paper was written to form part of a special symposium devoted to body tissues under extreme loading and disease, we have pointed out those areas where our approach has contributed, or may contribute in the future, to the understanding of bone mechanics under these particular conditions. However it should be emphasised that we have still a lot to do in the future in considering these particular aspects in more detail.

REPAIR OF BONE MICRODAMAGE: DOES IT MATTER?

Before considering the detection and repair of small cracks in bone – which is the subject of this paper – it is worth asking: "Does it matter?". Do these cracks threaten the structural integrity of bone in normal use? This is a question which we can answer using a different model which we have developed over the last 12 years, a phenomenological model which describes fatigue damage and repair in bone using the available data, which are now quite extensive, and including statistical scatter via the Weibull approach. This model has been described in a number of previous publications [6-9]; it does not consider at all the underlying physical mechanisms, but it is able to predict experimental data from different animals and different test protocols rather well, it gives reasonable predictions of phenomena such as the effect of exercise on stress fracture risk and as a result has recently has been employed in the field of sports science [10].

Here we use the model to consider the risk of fatigue failure in a typical bone, loaded with physiologically normal stresses, comparing bone from a typical individual capable of repairing small cracks as they initiate, and an individual suffering from a disease state in which this repair process does not occur. Figure 1 below presents the results in terms of a stress reduction factor, defined as the ratio between the stresses in the normal person and in the diseased person which give the same fatigue behaviour. The behaviour chosen was a 1% risk of fatigue failure per bone during the lifetime of the person, as this is typical of the actual rates of failure of bones in primates in the wild [11]. The stress factor is thus the same as the factor by which the compromised individual would have to reduce the forces on their bones in order to prevent failure from occurring.

Obviously the magnitude of this factor depends on the activity level of the individual: more active individuals will be more at risk, especially those engaging in extreme loadings such as professional athletes and soldiers. We considered six different activity levels as defined by Whalen et al [12]. The results show that repair has a very strong influence: for normal or active individuals it's necessary to reduce loadings by a factor of the order of 2 to 3 if repair is not available. In practice this would mean that any activity more strenuous than slow walking would be dangerous.

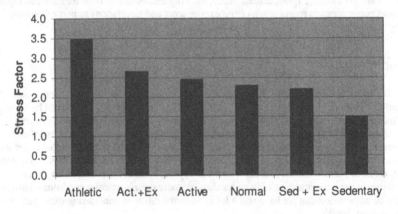

Figure 1: Estimated stress factor for similar fatigue performance of bones, comparing normal individuals and those having no ability to repair microcracks. Six different activity levels are considered, after Whalen et al [12].

MICROCRACK DETECTION: THE SCISSORS MODEL

Many workers have contributed to what is now quite an extensive research activity on the initiation, growth and repair of microcracks in bone. A recent review provides a summary of this work [13]. In brief, cyclic loading *in vivo* causes cracks to initiate in cortical and cancellous bone. These cracks are typically elliptical in shape, being elongated in the direction approximately parallel to the bone's longitudinal axis due to the material's anisotropy. On average the length (from tip to tip) of such a crack is about 100μm on its minor axis and about 400μm on the major axis. We will refer to the minor axis length in what follows because this is the length which appears on transverse sections of bones and so is the value most often reported as the crack length. . The number density of cracks varies greatly but can be in excess of 1 crack per cubic millimeter. *In vivo* crack lengths rarely exceed 200μm, but in cyclic loading tests conducted on bones *ex vivo*, a proportion (10-20%) of cracks are found to propagate by fatigue mechanisms, one crack eventually causing failure. Based on our experimental experience and some simple fracture mechanics calculations, one can form a picture of the risk which these cracks pose. Very small cracks (say less than 30μm) pose a negligible risk; cracks of the typical length of 100μm can be tolerated but would pose a risk in individuals engaging in strenuous activities; longer cracks (greater than about 300μm) will grow quite quickly under normal *in vivo* stresses and so must be detected and removed to avoid failure. Thus we have a kind of "specification" for the maintenance system which operates in our bones.

The repair system is quite well described: it consists of two types of cells working together: osteoclasts remove old, damaged bone and osteoblasts fill in the gap by making new bone. Currently though, we don't understand how microcracks are detected in the first place. There are a number of theoretical models under investigation (see [13]) most of which work from the idea that the crack affects the flow of fluid which is constantly seeping through the porous bone matrix. Our model is quite different: we considered whether the crack would damage any of the cells (osteocytes) which live in bone. We concluded that the cells themselves were unlikely to be fractured by the locally elevated strains near the crack, but that fracture could occur in the long, thin extensions known as cellular processes, which link cell to cell in a network known as the "syncytium". Each cell has about 100 such processes (see fig.2). We proposed that where a process passes across a crack, the displacements of the crack faces could cause rupture (fig.2). Theoretical analysis using fracture mechanics found that this was most likely to occur due to shear loadings, especially in combination with compressive forces across the crack faces. In fact most cracks in bone experience this combination of shear and compression, due to the typical loadings and crack orientations which occur.

EXPERIMENTAL WORK: MICROSCOPY

The photograph in fig.2 represents the highest level of resolution which we were able to achieve using laser confocal microscopy. To obtain more precise information we examined cracks at higher magnification using scanning electron microscopy. We found fibrous features spanning crack faces: by staining the samples with Phalloidin to reveal the cellular material, we demonstrated that the great majority of these were cellular processes. This is interesting in itself because other workers have proposed that these features are collagen fibrils and have argued that they have a significant toughening effect. For the bone samples which we examined, at least, there were very few collagen fibrils spanning cracks: this finding could be important because cellular processes would be too weak to contribute any significant toughening effect.

We first examined ovine bones which had been taken from animals used in other experiments, which were known to have had only normal loading activities during life. We saw both intact and broken processes across crack faces; we counted the intact ones because it was difficult to count the broken ones reliably.

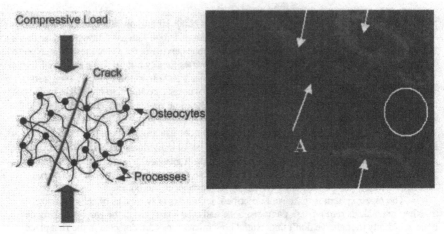

Figure 2: Schematic illustrating the Scissors Model. Laser scanning confocal image of a bone sample (width of picture = 50µm). Three osteocyte cells can be seen (arrowed) with many processes coming from them. A crack (arrowed, A) crosses the picture: some cellular processes cross the crack and remain intact (circled) whilst others have ruptured.

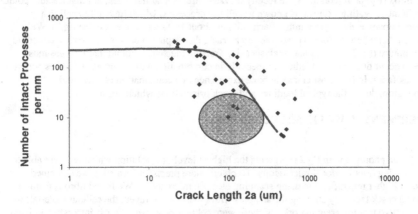

Figure 3: Experimental data (points) recording the number of intact processes per mm of crack length, as seen on transverse sections of ovine bone after normal *in vivo* loading. Predictions of our theoretical model (line). Recent data from samples subjected to loadings greater than *in vivo* levels (grey circle).

Figure 3 shows the results, plotting the number of broken processes per millimeter of crack length, as a function of crack length. The figure also shows the predictions of our theoretical model, using a typical osteocyte density (20,000/mm^3) and assuming normal *in vivo* loading activities. There is a lot of scatter in the data points from individual cracks, but it's clear that longer cracks have relatively fewer unbroken processes and that this behaviour is well predicted by the model. Figure 3 also indicates some more recent data obtained from ovine bone samples which had been subjected to cyclic loading in the laboratory at stress levels greater than experienced *in vivo*. This data, currently incomplete, is indicated by the grey circle, showing that for crack lengths around 100µm there are many more broken processes. All of this is nicely in accordance with our "specification" above: for *in vivo* loading there is a threshold at somewhat less than 100µm, below which no processes break and so the crack would be undetectable. This threshold shifts downwards if higher stresses are applied which would make the 100µm cracks potentially much more dangerous.

EXPERIMENTAL WORK: *IN VITRO* CELL STUDIES

In parallel with the above work we have been conducting studies on living cells cultured *in vitro*. We used MLO-Y4 cells, a line of cells developed for laboratory study, which have been shown to have many properties similar to those of natural osteocytes, but which are easier to culture and maintain for experimental work. We cultured these cells in 3D gels in pots of diameter 10mm, depth 10mm. A crack-like planar defect was introduced, width 5mm, depth 10mm, of varying thickness, using wires of different diameters. We made various measurements to detect the production of different proteins which are known to play roles in cell signalling. As figure 4 shows, we found that after introducing the cracks, there was a significant increase in the production of RANKL, a protein which is known to stimulate the production of osteoclasts. This is highly significant because osteoclast production is the first step in the repair process.

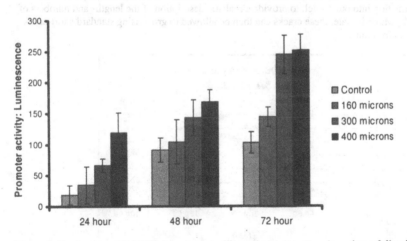

Figure 4: Production of RANKL by osteocyte-like cells *in vitro* at various times following introduction of a crack-like defect 5mm x 10mm, various widths (160-400µm), compared to undamaged controls.

These MLO-Y4 cells are similar to real osteocytes in bone, but there are some important differences, which we investigated experimentally. We found that real osteocytes have an average separation of 37μm and have typically 100 processes per cell. Our MLO-Y4 cells, on the other hand, had a separation of 100μm and only 13 processes per cell on average. This means that the "crack" created in our cell experiments is (in terms of the number of processes which cross it) equivalent to a much smaller crack in bone: in fact it is equivalent to a crack of length 500μm, somewhat larger than the typical 100μm, but more comparable than first appears. Given that our experimental cells are 100μm apart, the 160μm wire is probably cutting through processes but not rupturing cells (though obviously this is something we need to check) whilst the thicker wires have a greater effect because they probably are also breaking cell bodies, something which we believe does not happen in bone.

THEORETICAL MODEL

A theoretical model, written in the form of a MATLAB simulation, was first developed some years ago [14; 15], based around a fracture mechanics description of fatigue crack propagation in short and long cracks. This model is currently being improved to make several aspects more realistic. One feature which required attention was the simulation of crack initiation, which is a common problem in fracture mechanics models because we have no real theoretical framework for this stage of a crack's life. To circumvent this problem we examined the data on microcrack lengths in bone samples taken from our own work and other sources. We found that the distribution of crack lengths can be described as a two parameter Weibull distribution, whose constants change in a predictable manner with stress level, number of cycles, animal type etc. Figure 5 shows an example of one of these distributions. We intend to incorporate this information into our model, to provide a realistic description of the lengths and numbers of cracks which initiate; these cracks can then be allowed to grow using standard short-crack growth formulae.

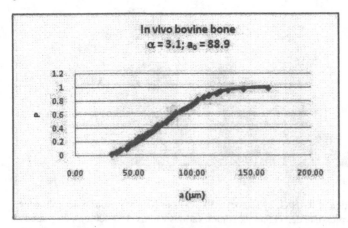

Figure 5: Example of our analysis of microcrack data, showing a two-parameter Weibull fit to the crack length distribution in bovine bone samples subjected to normal *in vivo* loadings.

DISCUSSION

As mentioned above, this paper is very much a description of work in progress rather than of a finished project. Results to date are very encouraging, suggesting that several aspects of our scissors model can be demonstrated in experiments and predicted theoretically. For example, we have now shown conclusively that cellular processes can span cracks, remaining intact in small cracks under *in vivo* loadings and breaking in increasingly large numbers if the crack length or the stress increases. Through our cell experiments we have demonstrated that planar defects of approximately the same size as microcracks (when differences in the cell types are accounted for) cause cells to increase production of a substance which stimulates bone-resorbing cells. We aim to incorporate all of these effects into our theoretical model, thus building a simulation of the entire feedback loop whereby microdamage is detected and repaired. One important and difficult step in the development of that model is the simulation of crack initiation: this problem has recently been overcome by studying and modeling crack length data.

What consequences does this work have for tissues under extreme loading and disease? High levels of loading, equivalent to very strenuous or active lifestyles, are known to give rise to stress fractures. This is a major problem, for example, in military recruits, football players and racehorses. Stress fractures occur because fatigue cracks are growing so quickly that the time between detection and repair is no longer sufficient to prevent the development of a macrocrack. Stress fractures can also happen as a result of diseases such as osteoporosis, which may affect the detection and repair system and also the mechanical properties of the bone itself. Drug treatments for osteoporosis work in different ways, targeting different parts of the feedback system. In all of these cases an understanding of the mechanisms of crack detection and repair is clearly vital.

Further work is required to confirm the scissors mechanism and to investigate it in more detail; this work is ongoing at present.

CONCLUSIONS

1. The repair of microdamage by living bone is vital to maintain the necessary level of structural integrity which allows us to go about everyday life without suffering fatigue failures.

2. The scissors model, in which cracks are detected because they cause rupture of cellular processes, is demonstrated to be a plausible mechanism.

3. In particular, the numbers of broken processes rise significantly when crack lengths increase and when stresses increase, within and just above normal physiological levels.

4. In addition, crack-like defects introduced into cell cultures cause increased production of RANKL, which is known to stimulate osteoclast production.

5. Microcracks in bone have a length distribution which can be described by the two-parameter Weibull equation: this finding is useful for the simulation of fatigue crack initiation in our theoretical model.

ACKNOWLEDGMENTS

This work was funded by Science Foundation Ireland as part of the Research Frontiers Programme, grant number 08-FRP-ENM991. We are grateful to Prof.Bonewald for permission to use the MLO-Y4 cell line, to Teuvo Hentunen and colleagues for advice concerning the cell culture experiments, and to David Burr and colleagues for generously sharing their microcrack data with us.

REFERENCES

1. Taylor, D., Hazenberg, J. G., and Lee, T. C., "The cellular transducer in damage-stimulated bone remodelling: A theoretical investigation using fracture mechanics," *Journal of Theoretical Biology*, Vol. 225, No. 1, 2003, pp. 65-75.

2. Hazenberg, J. G., Freeley, M., Foran, E., Lee, T. C., and Taylor, D., "Microdamage: a cell transducing mechanism based on ruptured osteocyte processes," *Journal of Biomechanics*, Vol. 39, 2006, 2096-2103.

3. Hazenberg, J. G., Taylor, D., and Lee, T. C., "The role of osteocytes in functional bone adaptation," *BoneKey Osteovision*, Vol. 3, 2006, pp. 10-16.

4. Hazenberg, J. G., Taylor, D., and Lee, T. C., "The role of osteocytes in preventing osteoporotic fractures," *Osteoporosis International*, Vol. 18, 2007, pp. 1-8.

5. Hazenberg, J. G., Hentunen, T., Heino, T. J., Kurata, K., Lee, T. C., and Taylor, D., "Microdamage detection and repair in bone: fracture mechanics, histology, cell biology," *Technology and Health Care*, Vol. 17, 2009, pp. 67-75.

6. Taylor, D. and Kuiper, J. H., "The prediction of stress fractures using a 'stressed volume' concept," *Journal of Orthopaedic Research*, Vol. 19, No. 5, 2001, pp. 919-926.

7. Taylor, D., "Scaling effects in the fatigue strength of bones from different animals," *Journal of Theoretical Biology*, Vol. 206, No. 2, 2000, pp. 299-306.

8. Taylor, D., "Fatigue of bone and bones: An analysis based on stressed volume," *Journal of Orthopaedic Research*, Vol. 16, No. 2, 1998, pp. 163-169.

9. Taylor, D., Casolari, E., and Bignardi, C., "Predicting stress fractures using a probabilistic model of damage, repair and adaptation," *Journal of Orthopaedic Research*, Vol. 22, 2004, pp. 487-494.

10. Edwards, W. B., Taylor, D., Rudolphi, T. J., Gillette, J. C., and Derrick, T. R., "Effects of stride length and running mileage on a probabilistic stress fracture model," *Medicine and Science in Sports and Exercise*, Vol. 41, No. 12, 2009, pp. 2177-2184.

11. Currey, J., *Bones: Structure and Mechanics*, Princeton University Press, USA 2002.

12. Whalen, R. T., Carter, D. R., and Steele, C. R., "Influence of physical activity on the regulation of bone density," *Journal of Biomechanics*, Vol. 21, No. 10, 1988, pp. 825-837.

13. Taylor, D., Hazenberg, J. G., and Lee, T. C., "Living with Cracks: Damage and Repair in Human Bone," *Nature Materials*, Vol. 2, 2007, pp. 263-268.

14. Taylor, D. and Lee, T. C., "Microdamage and mechanical behaviour: Predicting failure and remodelling in compact bone," *Journal of Anatomy*, Vol. 203, No. 2, 2003, pp. 203-211.

15. Taylor, D. and Lee, T. C., "A crack growth model for the simulation of fatigue in bone," *International Journal of Fatigue*, Vol. 25, No. 5, 2003, pp. 387-395.

Mater. Res. Soc. Symp. Proc. Vol. 1274 © 2010 Materials Research Society   1274-QQ08-07

Mechanical testing of limpet teeth micro-beams using FIB

Dun Lu[1] and Asa H. Barber[1]
[1]Department of Materials, School of Engineering and Materials Science, Queen Mary University of London, Mile End Road, London E1 4NS, United Kingdom

ABSTRACT

Discrete volumes of material in the form of a beam have been isolated from a parent limpet tooth using Focused Ion Beam (FIB) techniques. Mechanical bending tests of individual beams are performed using atomic force microscopy (AFM). The relatively small volumes tested in this beam-bending configuration allow approximation of the limpet tooth structure to a uniaxial short fibre composite. This mechanical testing technique is superior to conventional micro-hardness indentation as a defined stress condition within a locally defined volume is examined. Composite theory is shown to be valid for describing the mechanical properties of limpet teeth at sub-micron lengths scales and used to determine the synergy between the nanomaterial constituents.

INTRODUCTION

Limpet teeth perform a mechanical function for feeding and can be considered as a biological composite where mineral crystals of an iron oxide-hydroxide (Goethite) are expected to behave as a reinforcing phase for a polymeric chitin matrix. The mineral forms regular fiber-like crystals, often with a diameter of a few tens of nanometers [1], which show a complex organization within the tooth geometry. However, at micron to sub-micron length scales the mineral fibers are highly aligned and approximate to a uniaxial short fiber composite. Mechanical studies of limpet teeth at these length scales present an opportunity to elucidate behavior in a structural biological composite and ascertain the influence of the nanomaterial constituents on resultant mechanical performance.

The capability of Focused Ion Beam (FIB) to be used as a milling tool for producing micrometer-sized samples for mechanical testing has begun to attract increasing attention in the study of synthetic materials [2]. FIB has also proven suitable for site-specific milling of subsurface structures of biological samples for conventional electron microscopy examination [3]. The ability to select specific volumes using FIB is particularly important for the understanding the mechanical properties of limpet teeth material constituents, which is not possible at larger length scales where the shape and geometry of limpet tooth in addition to the inherent material properties define mechanical behavior of the tooth.

The characterization of mechanical properties of materials at the sub-micro scale is typically carried out using atomic force microscopy (AFM) in contact mode. Mechanical testing using AFM has been previous shown to be suitable for measuring the mechanical properties of a number of nanomaterials and their composites [4, 5, 6]. Mechanical testing on limpet teeth is sparse, despite its potential as a high performance biological fibrous nanocomposite, but some work has examined macrohardness behavior [7]. While these macroscopic experiments are useful in understanding the overall mechanical properties of limpet teeth, currently there is little understanding on how the nanoscale constituents define overall mechanical behavior. In this work we use AFM to measure the bending behavior of limpet teeth micron sized beams prepared by FIB.

EXPERIMENT

Fabrication of micro-cantilever beams

Mature limpet teeth harvested from the limpet patella vulgata were mounted onto a standard scanning electron microscope (SEM) aluminum sample-mounting stub using a silver paint adhesive. Micron sized beams were then milled in the samples using a FIB combined with an imaging SEM in a dual-beam system (FEI Quanta 3D, FEI, EU/USA). FIB milling was conducted using gallium ions accelerated at 30kV and a beam current of 1nA. A suitable region in the middle of the tooth sample was first located and beams FIB milled so that the long axis of the beams were parallel to the long axis of the tooth as shown in Figure 1.

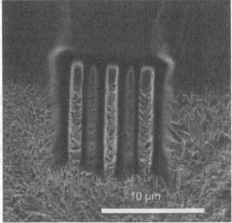

Figure 1. SEM micrograph showing rectangular cross-section beams FIB milled at the surface of the limpet tooth

The dimensions of the FIB milled beams were typically 10μm x 2μm x 1μm. An array of beams as shown in Figure 1 was fabricated from the cusp of the limpet tooth towards the base of the limpet tooth as shown in Figure 2. As the mineral fibers in limpet teeth are parallel to the surface of the tooth, the long axis of the FIB milled beams were varied relative to the long axis of the tooth in order to examine if subsequent mechanical testing could elucidate the difference in mineral fiber orientation.

114

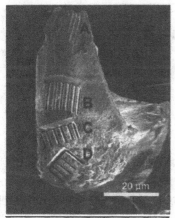

Figure 2. SEM micrograph showing four FIB milled beam regions at the surface of the limpet tooth

Micro-cantilever beam bending measurement

The limpet tooth incorporating the FIB milled beam arrays as shown in Figure 2 was transferred to a conventional AFM system (NTegra, NT-MDT, Rus.) in order to perform mechanical testing of each individual beam. Micromechanical measurements were carried out by contacting the end of the AFM probe with the end of the FIB milled limpet tooth beam. An applied force caused the limpet tooth beam to bend, with the applied force and beam deflection recorded using the AFM. Specifically, the AFM probe was attached to a cantilever-based system where cantilever deflection, corresponding to applied force, could be monitored using a photodiode. The sample was moved up to the AFM probe using a piezo-positioner until contact of the beam with the AFM probe and subsequent beam bending occurred. The limpet tooth beam deflection was obtained by subtracting the AFM cantilever deflection from the piezo-positioner displacement.

RESULTS AND DISCUSSION

Limpet tooth beam bending tests from four regions on the limpet tooth surface, shown as A-D in Figure 2, gave a series of beam stress-beam deflection curves as shown in Figure 3.

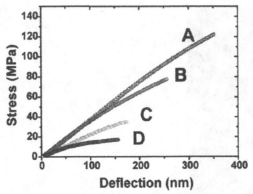

Figure 3. Stress-limpet tooth beam bending deflection curves corresponding to four regions defined in Figure 2.

The apparent elastic modulus of each beam was calculated using Euler–Bernoulli beam bending theory:

$$\frac{\sigma}{\delta} = \frac{E.h^2}{4L^3}$$

Where σ and δ are the applied stress in the beam and beam deflection respectively, E is the elastic modulus of the beam, h is the thickness of the beam and L is the length of the beam. The elastic modulus of each beam is generally observed to decrease from region A to D. The elastic modulus plotted verses orientation of the long axis of the beam relative to the tooth long axis is shown in Figure 4 below. We observed that the elastic modulus of the beam measured from bending experiments varies from 140GPa when the beam is parallel to the tooth long axis to 60GPa when the beam is aligned 45°to the long axis. An average fit to the elastic modulus to the beam long axis shows a non-linear relationship shown as a fit line in Figure 4.

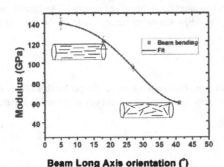

Figure 4. Plot of elastic modulus of FIB milled beams versus beam long axis orientation relative to the long axis of the limpet tooth.

116

The different elastic modulus of the limpet tooth beams is considered related to the mineral orientation within the tooth beams. Composite models describing the effects of reinforcing fiber orientation on the elastic modulus of composites has been previously described, most notably using the Halpin-Tsai model [8]. This model indicates that the elastic modulus of a fiber-reinforced composite is largest when the fibers are aligned parallel to the loading direction, as observed in region A. Increasing 'off-axis' fiber orientation as observed progressively from region A to D causes less strain in the fiber during beam bending and thus gives a corresponding drop in the reinforcing behavior of the fibrous material. Our results shown in Figure 4 suggest that fiber orientation effects described using composite theory are applicable to the limpet tooth materials.

CONCLUSIONS

Micron sized beams have been FIB milled from the surface of a limpet tooth and subsequently mechanically tested using AFM. Calculated elastic modulus values for the limpet tooth beams indicate that mineral orientation within the beam defines the overall mechanical properties of the beam. Thus, the mechanical behavior of discrete volumes defined by FIB milling values can be explained by established composite ideas.

ACKNOWLEDGMENTS

The authors would like to thank the NanoVision Centre at Queen Mary University of London for the use of facilities. We are grateful for financial support from the Engineering and Physical Science Research Council, UK (grant award EP/E039928/1).

REFERENCES

1. S. Mann, C. C. Perry, J. Webb, B. Luke, R. J. P. Williams, Proc. R. Soc. Lond. B 227, (1986), 179–190
2. D. Di Maio, S.G. Roberts, Mater. Res. J. 20, (2005), 299–302
3. D. Drobne, M. Milani, V. Lescaroner, F. Tatti,. Microscopy Res. and Technique 70, (2007), 895–903.
4. W. Wang, T. Peijs, A. H. Barber, Nanotechnology 21, (2010), 035705
5. W. Wang, A. J. Bushby, A. H. Barber, Appl. Phys. Lett. 93, (2008), 201907
6. W. Wang, P. Ciselli, E. Kuznetsov, T. Peijs, A. H. Barber, Phil. Trans. Royal. Soc. A 366, (2008), 1613-1626
7. P. van der Wal, H. J. Giesen, J. J. Videler, J. Mater. Sci. C 7 (2000), 129 – 142
8. J.C. Halpin, J. L. Kardos, Polymer Engineering and Science 16 (5), 344 (1976).

Mater. Res. Soc. Symp. Proc. Vol. 1274 © 2010 Materials Research Society   1274-QQ08-08

Experimental investigations into the mechanical properties of the collagen fibril-noncollagenous protein (NCP) interface in antler bone

Fei Hang[1], Asa H. Barber[1]
[1]Centre for Materials Research & School of Engineering and Materials Science, Queen Mary University of London,
Mile End Road, London E1 4NS, UK.

ABSTRACT

Antler is an extraordinary bone tissue that displays significant overall toughness when compared to other bone materials. The origin of this toughness is due to the complex interaction between the nanoscale constituents as well as structural hierarchy in the antler material. Of particular interest is the mechanical performance of the interface between the collagen fibrils and considerably smaller volume of non-collagenous protein (NCP) between these fibrils. This paper directly examines the mechanical properties of isolated volumes of antler using combined in situ atomic force microscopy (AFM)-scanning electron microscopy (SEM) experiments. The antler material at the nanoscale is approximated to a fiber reinforced composite, with composite theory used to evaluate the interfacial shear stresses generated between the individual collagen fibrils and NCP during mechanical loading.

INTRODUCTION

Bones, especially compact bone, is a stiff and tough material having the major functions of providing mechanical support and bearing external load. Bone has a structural organization over different length scales and constituent properties expected to contribute to mechanical performance [1-3]. Collagen fibrils are one such constituent that provide an organic framework and act as a template for the mineralization of carbonated apatite crystals. The irregular shaped apatite crystals are located both within collagen fibrils and extra-fibrillar regions. These discrete fibrils are organized over different hierarchical length scales to provide optimized mechanical function [1, 4-7]. Importantly, the spaces between the collagen fibrils are filled with predominantly with non-collagen protein (NCP). As with other fiber-reinforced composites, the interface between the fibrous material and surrounding NCP is expected to be critical in defining the overall mechanical properties of the material. Bone can therefore be considered as mineralized collagen fibrils acting as the reinforcement while NCP between fibrils is transfers applied load between the fibrils. Interfacial mechanics between fibers and surrounding polymer matrix has contributed to significant composite theory and is relevant at a number of different length scales [8-14]. The application of composite theory to the mechanical response of the mineralized collagen fibril-NPC interface is therefore critical in understanding the influence of nanoscale structure on the fracture behavior of bone. However, direct testing of this collagen fibril-NCP interface is particularly challenging due to the relatively small volumes considered and has yet to be determined.

Measurement of the interfacial mechanical properties within fiber reinforced composite is typically carried out using single fiber pullout testing, previously shown at a range of different scales [8-14]. In the single fiber pullout test, part of the fiber is embedded within the solid matrix material, typically a polymer. A force applied to the free length of the fiber causes an increase in the shear stress at the interface until, at a critical interfacial shear stress, the interface will fail as the adhesion between the fiber and matrix is overcome. The maximum force required to fail the interface and separate the fiber from matrix can be used to calculate the interfacial shear strength between the fiber and matrix. An approximation of the interfacial shear strength is often calculated by simply dividing the maximum force applied over the total fiber embedded area within the matrix [8-10]. However, while measuring the interfacial shear strength between engineering fibers with diameters of the order of microns has been extensively investigated, fibers with smaller diameters such as found in collagen fibrils from bone requires non-traditional mechanical testing techniques. The development of atomic force microscopy (AFM) as a nanoscale testing technique has been critical in developing mechanical understanding of composite materials using nanomaterial consitituents. The AFM is particularly beneficial for mechanical testing of nanomaterials due to its high resolution imaging and the capacity to applying and record relatively small forces using the AFM probe situated at the end of a flexible bending beam, known as the cantilever. AFM has been used extensively in measuring the mechanical properties of biological materials, with recent work highlighting the high force resolution of AFM while deforming individual tropocollagen macromolecules [15]. Pioneer work in recent years has extended the AFM to investigate the interfacial mechanics between carbon nanotubes reinforcing a polymer using individual carbon nanotube pullout experiments[8, 10]. This work extends the AFM nanomechanical testing of carbon nanotube-polymer interfaces to the collagen fibril-NCP interface found in bone. We note that while mechanical testing on individual collagen fibrils has been carried out [16] [17], only recent work has individual mineralized collagen fibrils until failure from antler bone using a novel combination of AFM and SEM [18]. In this work we use the same combined AFM and SEM setup to evaluate the shear strength of fibril-NPC interface by pulling out the individual collagen from the bone tissue matrix.

EXPERIMENT

A custom built AFM (Attocube GmbH, Ger) was fitted within the chamber of a scanning electron microscope (SEM, FEI Company, EU/USA). The SEM chamber pressure was maintained from $1x10^{-3}$ to $5x10^{-3}$ Pa during the pullout of collagen fibrils from a parent sample of antler. The combination of both SEM and AFM is powerful as the AFM provides high-resolution force information while the SEM gives imaging capabilities. Antler samples were cut from the main antler beam of a mature red deer (Cervus elaphus,). Small beams with dimensions of 3x20x0.2 mm were cut from the bulk material by using a water-cooling rotating diamond saw (Struers Accatom-5) with the long direction oriented parallel to the main beam direction (which is also the principal osteonal axis). Specimens were transferred into Hank's Balanced Solution and left for at least 24 hours to rehydrate the bone material fully and mitigate mineral loss that may occur in distilled water or physiological saline. Wet samples were fractured perpendicular to the long axis to expose mineralized collagen fibrils. After the removal of water on the sample surface using filter paper to avoid the interference on SEM imaging, specimen was transferred into the SEM chamber immediately.

The fracture surface of the antler bone material is shown in Figure 1. Individual fibrils were observed at the fracture surface using SEM secondary electron imaging at relatively long

120

working distances (15mm) and low accelerating voltages (2kV). These exposed fibrils were considered as potentially suitable candidates for a pullout test. A candidate fibril was examined at orientations within the SEM to ensure that the chosen collagen fibrils is embedded within the fracture surface rather than simply attached on the surface. The pullout test was performed by first submerging the end of the AFM probe into epoxy glue (Poxipol, Arg.) and then translating the AFM setup using xyz piezo-positioners until the end of the AFM probe containing the glue contacted an individual collagen fibril. A typically testing configuration is shown in Figure 1(a). Translating the AFM probe away from the collagen fibril along its principle axis caused increase in the shear stress at the fibril-NCP interface until the applied force was sufficient to cause interfacial failure and collagen fibril pullout, as shown in Figure 1(b). Force-displacement curves were produced from accurate measurement of the AFM cantilever deflection using an optical interferometer setup for the recording of the stress and calibrated piezoelectric positioner minus the cantilever deflection for the calculation of AFM probe displacement.

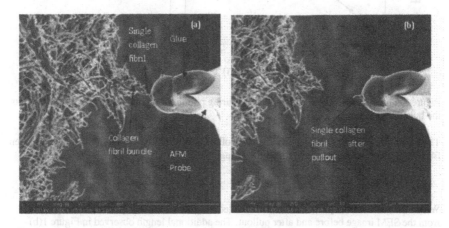

Figure 1. (a) SEM micrograph of an individual collagen fibril protruding from a collagen bundle attached to the end of an AFM probe. (b) The individual collagen fibril pulled-out from the collagen bundle achieved by translating the AFM probe away from the fracture surface.

DISCUSSION

Pullout testing of 5 separate mineralized collagen fibrils found in a fibrillar bundle at the fracture surface of antler bone was carried out, with a typically force-displacement curve shown in Figure 2. The plot in Figure 2 shows that the force applied to the collagen fibril increases with AFM probe displacement until, at a critical applied force, the interface fails and the force drops to zero. The plot shows that the progressive increase of force with displacement until pullout in Figure 2 is indicative of interfacial failure only. Fibril failure at some point within the bulk material followed by pullout of the fragment during the AFM applying force to the fibril does not occur as the force plot if Figure 2 would first show an increase in the force with displacement until, at fibril failure within the bulk, a drop in the force corresponding to fibril fracture. The force-displacement plot in Figure 2 therefore indicates that the exposed collagen fibrils have fragmented within the bulk, thus allowing pullout of a corresponding embedded length.

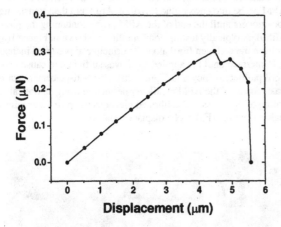

Figure 2. Force-displacement curve recorded during pullout of an individual collagen fibril from the bulk antler material.

The maximum pull-out force P applied to the collagen fibril can be used to measure the collagen fibril-NCP interfacial shear strength (IFSS) using:

$$\tau = P/\,\xi l_e \tag{1}$$

Where the ξ is the fiber perimeter. The embedded length l_e of the collagen fibril is calculated from the SEM image before and after pullout. The additional length observed in Figure 1(b) when compared to the free length of the fibrils in Figure 1(a) is therefore used as the embedded length. The diameter of the fibrils tested is 115.6±32.8 nm. The average IFSS calculated from 5 pullout tests using Equation 1 gives a collagen fibril-NCP interfacial shear strength of 6.2±1.5MPa. This strength is considerably smaller than the shear strength of the bulk materials of antler bone recorded by Chen et al.[4] and indicates that in shear testing on bulk antler bone causes cracks to propagate both along the relatively weak fibril-NCP interface but through the fibrils causing their fracture and an enhancement in the shear strength value.

In addition, many of the calcium apatite minerals in antler are intra-fibrillar and provide a model composite fiber system, as opposed to many other bones where the mineral is also found in extra-fibrillar spaces [4, 19-21]. Therefore we believe our results measure the shear strength behavior of the NCP only and not for NCP incorporating significant amounts of mineral.

CONCLUSIONS

Pullout test on individual mineralised collagen fibrils from antler bone matrix have been carried out using a combined AFM-SEM system. The SEM images provide the capability of observing manipulation and fix an individual collagen fibril to the end of an AFM probe. Forces applied to exposed collagen fibrils at a fractured antler surface were recorded by AFM and shown using a force-displacement curve. Average shear strength of the interface calculated from the maximum pullout force applied to the exposed collagen fibril gave values of 6.2±1.5MPa for 5 individual collagen fibril pullout tests. This strength is smaller than the shear strength of the bulk antler bone material, which we believe is indicative of bulk shear testing causing fibril fracture as well as interfacial failure.

ACKNOWLEDGMENTS

The authors would like to thank the NanoVision Centre at Queen Mary for the use of facilities. Financial support through the Engineering and Physical Science Research Council, UK (grant award EP/E039928/1) is gratefully acknowledged.

REFERENCES

[1]   J. D. Currey, "The design of mineralised hard tissues for their mechanical functions," Journal of Experimental Biology, vol. 202, pp. 3285-3294, Dec 1999.
[2]   S. Weiner, W. Traub, and H. D. Wagner, "Lamellar bone: Structure-function relations," Journal of Structural Biology, vol. 126, pp. 241-255, Jun 30 1999.
[3]   J. Y. Rho, L. Kuhn-Spearing, and P. Zioupos, "Mechanical properties and the hierarchical structure of bone," Medical Engineering & Physics, vol. 20, pp. 92-102, Mar 1998.
[4]   P. Y. Chen, A. G. Stokes, and J. McKittrick, "Comparison of the structure and mechanical properties of bovine femur bone and antler of the North American elk (Cervus elaphus canadensis)," Acta Biomaterialia, vol. 5, pp. 693-706, Feb 2009.
[5]   D. Vashishth, K. E. Tanner, and W. Bonfield, "Contribution, development and morphology of microcracking in cortical bone during crack propagation," Journal of Biomechanics, vol. 33, pp. 1169-1174, Sep 2000.
[6]   H. S. Gupta and P. Zioupos, "Fracture of bone tissue: The 'hows' and the 'whys'," Medical Engineering & Physics, vol. 30, pp. 1209-1226, Dec 2008.
[7]   K. Tai and C. Ortiz, "Nanomechanical heterogeneity of bone at the length scale of individual collagen fibrils," Abstracts of Papers of the American Chemical Society, vol. 231, pp. 179-PMSE, Mar 2006.
[8]   S. Nuriel, A. Katz, and H. D. Wagner, "Measuring fiber-matrix interfacial adhesion by means of a 'drag-out' micromechanical test," Composites Part a-Applied Science and Manufacturing, vol. 36, pp. 33-37, 2005.
[9]   C. Y. Yue and W. L. Cheung, "INTERFACIAL PROPERTIES OF FIBER-REINFORCED COMPOSITES," Journal of Materials Science, vol. 27, pp. 3843-3855, Jul 1992.

[10]   A. H. Barber, S. R. Cohen, S. Kenig, and H. D. Wagner, "Interfacial fracture energy measurements for multi-walled carbon nanotubes pulled from a polymer matrix," Composites Science and Technology, vol. 64, pp. 2283-2289, Nov 2004.

[11]   C. H. Hsueh, "INTERFACIAL DEBONDING AND FIBER PULL-OUT STRESSES OF FIBER-REINFORCED COMPOSITES .6. INTERPRETATION OF FIBER PULL-OUT CURVES," Materials Science and Engineering a-Structural Materials Properties Microstructure and Processing, vol. 149, pp. 11-18, Dec 1991.

[12]   J. K. Kim, C. Baillie, and Y. W. Mai, "INTERFACIAL DEBONDING AND FIBER PULL-OUT STRESSES .1. CRITICAL COMPARISON OF EXISTING THEORIES WITH EXPERIMENTS," Journal of Materials Science, vol. 27, pp. 3143-3154, Jun 1992.

[13]   A. H. Barber, S. R. Cohen, A. Eitan, L. S. Schadler, and H. D. Wagner, "Fracture transitions at a carbon-nanotube/polymer interface," Advanced Materials, vol. 18, pp. 83-87, Jan 2006.

[14]   R. A. Latour, J. Black, and B. Miller, "FRACTURE MECHANISMS OF THE FIBER MATRIX INTERFACIAL BOND IN FIBER-REINFORCED POLYMER COMPOSITES," Surface and Interface Analysis, vol. 17, pp. 477-484, Jun 1991.

[15]   T. Gutsmann, T. Hassenkam, J. A. Cutroni, and P. K. Hansma, "Sacrificial bonds in polymer brushes from rat tail tendon functioning as nanoscale velcro," Biophysical Journal, vol. 89, pp. 536-542, Jul 2005.

[16]   J. S. Graham, A. N. Vomund, C. L. Phillips, and M. Grandbois, "Structural changes in human type I collagen fibrils investigated by force spectroscopy," Experimental Cell Research, vol. 299, pp. 335-342, Oct 2004.

[17]   J. A. J. van der Rijt, K. O. van der Werf, M. L. Bennink, P. J. Dijkstra, and J. Feijen, "Micromechanical testing of individual collagen fibrils," Macromolecular Bioscience, vol. 6, pp. 697-702, Sep 2006.

[18]   F. Hang, D. Lu, and A. H. Barber, "Combined AFM-SEM for mechanical testing of fibrous biological materials," in Structure-Property Relationships in Biomineralized and Biomimetic Composites. vol. 1187, D. Kisailus, L. Estroff, H. S. Gupta, W. J. landis, and P. D. Zavattieri, Eds. Warrendale: Materials Research Society, 2009, pp. 135-140.

[19]   J. D. Currey, K. Brear, and P. Zioupos, "Dependence of Mechanical-Properties on Fiber Angle in Narwhal Tusk, a Highly Oriented Biological Composite," Journal of Biomechanics, vol. 27, pp. 885-&, Jul 1994.

[20]   D. Vashishth, J. C. Behiri, and W. Bonfield, "Crack growth resistance in cortical bone: Concept of microcrack toughening," Journal of Biomechanics, vol. 30, pp. 763-769, Aug 1997.

[21]   P. Zioupos, X. T. Wang, and J. D. Currey, "The accumulation of fatigue microdamage in human cortical bone of two different ages in vitro," Clinical Biomechanics, vol. 11, pp. 365-375, Oct 1996.

Mater. Res. Soc. Symp. Proc. Vol. 1274 © 2010 Materials Research Society   1274-QQ08-10

in vitro Study of a Novel Method to Repair Human Enamel

Song Yun[1], Yanjun Ge[2], Yujing Yin[1], Hailan Feng[2], Haifeng Chen[1,*]
[1] Department of Biomedical Engineering, College of Engineering, Peking University, Beijing 100871, China
[2] Department of Prosthodontic, Peking University School and Hospital of Stomatology, Beijing, 100081 China
* Corresponding author.

ABSTRACT

The chemical regeneration method of human enamel was improved for the probable clinical application: an oral medical device was designed to significantly reduce the necessary amount of remineralization liquid; the reactive liquid phase system was redesigned, thus the solution could be stored for long time.

INTRODUCTION

Dental enamel is the outmost layer of the human tooth [1]. Since there are no living cells in mature enamel, once been damaged, it cannot be self-repaired. In our previous work, we succeeded to synthesize human-enamel-like FA structures through liquid deposition [2]. This method provided a new approach to repair teeth with shallow surface caries, or other types of enamel defect. However, this method demands mass reactive liquid to soak the teeth *in vitro*, while the liquid can hardly be storage for long term (>1d). Therefore, it cannot be directly applied in clinical practice.

In order to gain the clinical maneuverability of this method, we improved it on two main aspects. At first, the pretreatment step was modified and the reactive liquid phase system was redesigned. An oral medical device to assist the reconstruction process was designed with its detailed operating steps.

EXPERIMENT

Instruments and materials

At first a polymer hollow hemisphere shell of about 4-5mm in diameter and 2mm in height, with two circular hole of 2mm interval was prepared (Fig.1). The liquid phase was divided to two separated reactive solutions – Solution A [contents 0.20M HEDTA, 0.20M Ca(NO$_3$)$_2$, 0.62M KOH] and Solution B [contents KH$_2$PO$_4$ 0.12M, KF 0.04M]. These two types of solution were chemical stable and could be kept individually for more than one month. While in use, mixing A and B would generate a chemical solution at pH 6.0 which contents 0.10M HEDTA-Ca, 0.06M H$_2$PO$_4^-$ and 0.02M F$^-$, the same as the reaction conditions of the previous liquid deposition method.

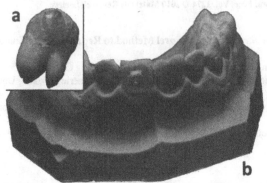

Figure 1. The polymer hollow hemisphere shell repair device on an *in vitro* tooth (a) and on an oral cavity model (b)

The demand quantity of reactive solution

The demand quantity of reactive solution, or the minimum volume of the hollow hemisphere device, was calculated by the following processes. The mass of the closely arranged deposited fluorapatite (FA) crystals, m_c, the density of the crystal, d_c, and the volume of the deposited crystal film, V_c, (which was equal to the product of the area of enamel base for them to deposit, S, and the average height of the crystals, h_c) obeyed the following relational expression,

$$m=d_c \cdot V_c = d_c \cdot S \cdot h_c \qquad (1)$$

Meanwhile, the crystal mass was also related with the quantity of reactive solution in the aqueous system as the following relational expression,

$$m=M \cdot n=m \cdot c \cdot V_l = M \cdot c \cdot S \cdot h_l \qquad (2)$$

where M represents the molar mass of FA ($504.3\ \mathrm{g \cdot mol^{-1}}$); n, the amount of FA; c, the resultant concentration of FA deposition; V_l, the volume of reactive solution; h_l, the average height of the device inner space to content the solution.

Since the FA crystal deposited on the same base area as where the solution was stored, the S in relational expression (1) and (2) could be reduced, and the ratio of h_l/h_c could be expressed as,

$$h_l/h_c = d_c/(M \cdot c) \qquad (3)$$

After substitute the value of d_c $3.18 \times 10^3\ \mathrm{kg \cdot m^{-3}}$ [3], M $504.3 \times 10^{-3}\ \mathrm{kg \cdot mol^{-1}}$ and c $0.02 \times 10^3\ \mathrm{mol \cdot m^{-3}}$, the value of h_l/h_c could be calculated 315.3. Based on previous experience, the height of the mature deposited FA crystal was approximately 4~5 µm. The average inner height of the hollow hemisphere shell should be 1.3mm at minimum. The area enclosed by the hemisphere shell was 8-10 mm^2, therefore its capacity should be 10-13 µL at minimum.

Evaluate the effectiveness of the new reactive liquid phase system

The effectiveness of the new liquid phase system was evaluated by the previous liquid deposition method. 100mL Solution A and 100mL Solution B was mixed and kept in a near-physiological environment. A pretreated (disinfected, polished, and etched by phosphoric acid) excised human tooth was immersed in it for 4 days. Then its surface feature was investigated by scanning emission microscopy (SEM).

Use the oral device to repair enamel

In the beginning of the new repair process, a drop of H_2O_2 solution was distributed and kept on the labial surface of disinfected human tooth for 1min. The pretreated tooth should be treated through 8 steps to be repaired as the list below (Fig2). First of all, two little pieces of photo-curing sealant was laid on the edge of the area to be repaired. The hollow hemisphere shell was put on this area, edge contacting with the sealant. The sealant should be illuminated for 30s to become stabilized. In the following steps, more sealant was laid on the whole edge of the hemisphere, and was illuminate it for another 30s, thus the hemisphere was toughly banded to the tooth surface. And then distilled water was injected to the hemisphere to measure its practical capacity, and to test its sealing quality. After the water was drained out, isopyknic Solution A and B (approximately 5-15µL for each of them, based on the specific surface shape of the tooth. The total solution amount should more than the minimum limit of 10-13µL according to the calculation result suggested above to ensure that the reaction material was sufficient) was injected to the inner space of the hemisphere. After the hemisphere shell was full-filled with the mixed solution, the two holes on the top of the hemisphere should be sealed with wax. Thereafter the tooth was put into an oral-cavity-like environment (37°C, 1atm, and saturated steam) for 1-4 days. When the main repair process was completed, the hemisphere shell and the photo-curing sealant was removed.

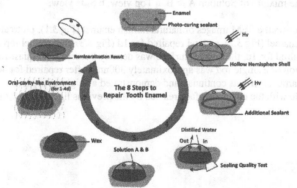

Figure 2. The 8 operation steps to repair human enamel. 1 lay two sealant spots. 2 put the hemisphere device on. 3 seal the device. 4 inject water to measure the capacity and to test the sealing quality. 5 inject Solution A & B. 6 close the top two hole with wax. 7 store it in an oral-cavity-like environment. 8 remove the device and the sealant.

RESULTS and DISCUSSION

Figure 3 shows the SEM images of the FA film deposited on the excised human tooth, which had been treated by the previous liquid deposition method with the mixture of Solution A & B. The deposited crystals have typical apatite hexagonal cross sections of approximately 300 nm in diameter and 5μm in height, which proved the effectiveness of this system to regenerate the tooth enamel.

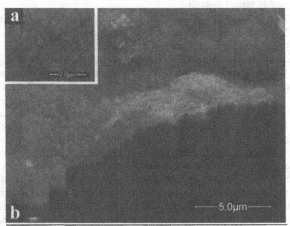

Figure 3. The SEM images of human enamel surface treated by the previous liquid deposition method with the mixture of Solution A & B. a Top view. b Side view.

Figure 4 shows the SEM images of natural human enamel (Fig 3.1), pretreated (30% H_2O_2, 1min) human enamel (Fig 3.2), enamel repaired for 1d (Fig 3.3) and enamel repaired for 4d (Fig 3.4). After repaired for 1d, the enamel surface was densely covered with quasi-rod structures. The cross-section of a single rod was approximately 300nm. After repaired for 4d, the regenerated enamel layer was continual, and demonstrated a height of approximately 5μm, which was comparable with the enamel surface reacted by the previous liquid deposition method for 12h.

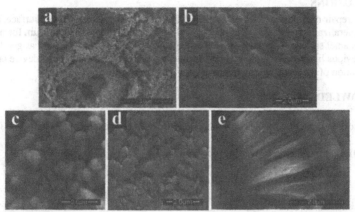

Figure 4. The SEM images to demonstrate the remineralization effect. a. natural human enamel. b pretreated (by 30%H2O2) human enamel. c *in vitro* repaired for 1 day. d&e in-vitro repaired for 4 days.

There are three possible explanations to explain the irregularity of the generated crystal on the repaired area. In the beginning of the reaction process, kinetically favored amorphous fluoridated calcium phosphate (AFCP) forms initially [4]. The AFCP micro-particles generate on, or generate in the liquid phase and would be captured on the enamel base, and then transform to the thermodynamically stable FA crystal. This process may need a superfluous amount of AFCP to sufficiently fill the enamel base. However, the amount of the reactant in the hemisphere shell device is exactly, or is only a little more than the necessary amount for the stoichiometry. Thus some area of the enamel base may has been naked to form some gaps among the final crystal product. Therefore the crystals cannot be regularly arranged.

Besides the insufficiency of AFCP, the homogeneous nucleation process may consume some reactant, which leads to the insufficiency of the reactant, thus the product. During the nucleation process, there may be a chemical balance between the homogeneous nucleation and the heterogeneous nucleation. Some homogeneous nucleus may also grow into mature crystals and consume some of the reactant.

Another explanation is related to the possible physical leakage of the reactant. Some reactant may seep through the interval among the natural enamel crystal into the tooth tissue to be consumed.

The sealing quality of the hemisphere device is very important for this new repairing method. Since it needs very little solution (only 10-30μL), slight leakage will seriously reduce the concentration of the reacting system, and the remineralization result. Besides, leaked hydroxyethyl ethylenediamine triacetic acid (HEDTA) is harmful to human body [5]. A common leaking location is the boundary between enamel and cementum, through which the reacting solution may seep into the tissue of the tooth. Therefore the hemisphere shell should not be laid across that boundary.

CONCLUSIONS

This repair method can work on a specific area of defeated human enamel surface, after 1-4 day remineralization process, a continual crystal film of about approximately 5μm forms on the natural enamel surface. The shape and the arrangement of these crystals are not as good as those using previous liquid deposition methods. Increasing the liquid capacity of the device or the concentration of the reactant may improve the quality of repair.

ACKNOWLEDGMENTS

The work is supported by the National Natural Science Foundation of China (NSFC Grant

50702001) and the Peking University 985-II grants.

REFERENCES

1. A. Boyde, in Teeth, ed. B. K. B. Berkovitz, A. Boyde, R. M. Frank, H. J. Hoehling, B. J. Moxham, J. Nalbandian and C. H. Tonge, *Handbook of Microscopic Anatomy*. Springer, Berlin, Germany (1988), vol. 5, pp. 309–473.
2. Y. Yin, S. Yun, J. Fang and H. Chen, *Chemical Communications*. **39**, 5892-5894 (2009).
3. Palache, C., H. Berman, and C. Frondel, *Dana's system of mineralogy* (7th edition). v. II, 879–889 (1951).
4. H. Chen, Z. Tang, J. Liu, K. Sun, S. R. Chang, M. C. Peters, J. F. Mansfield, A. Czajka-Jakubowska, B. H. Clarkson, *Advanced. Materials*. **18**, 1846-1851 (2006).
5. Anon, *International Journal of Toxicology*. **21**, 5, 95-142, Suppl. 2 (2002).

Multiscale Soft Tissue Mechanics

Mater. Res. Soc. Symp. Proc. Vol. 1274 © 2010 Materials Research Society   1274-QQ09-10

Surgical adhesive/soft tissue adhesion measured by pressurized blister test

Muriel L. Braccini[1], Bertrand R.M. Perrin[2], Cécile Bidan[1], Michel Dupeux[1]
[1]SIMaP - CNRS/Grenoble-INP/ UJF, BP 75, F38402 Grenoble, FRANCE
[2]C.H.U. Grenoble, FRANCE.

ABSTRACT

The practical adhesion of equine pericardium membranes bonded with surgical glue has been measured by the bulge-and-blister technique under injection of pressurized distilled water. The value of the interfacial crack propagation energy can be estimated from the critical debonding pressure. The measured practical adhesion energies are weak with regards to those of engineering structural adhesives, but they are reliable enough to allow a comparison between different surgical glues and a study of the influence of the bonding experimental conditions.

INTRODUCTION

Surgical adhesives are a very exciting and a promising area for development. They are increasingly used for skin suture, to reinforce anastomotic integrity in vascular repair, or as a sealant for closing gas or liquid leak [1-2]. Many different surgical adhesives have been developed over the past 30 years, varying in composition according to their surgical specialty: cyanoacrylate-based adhesive, fibrin glue or two components sealants with aldehydes as activators. All of them have to fulfill several criteria, among which bio-compatibility and adhesion to tissue. Bio-compatibility is well established by standards like ISO 10993, but there is no equivalent standard for surgical adhesive efficacy. Many clinical studies have been performed to evaluate this efficacy but they do not provide reliable quantitative information on the strength of the seal. Some studies on surgical glue are based on mechanical testing using the single lap shear test [3-5], which is commonly used to determine the strength of industrial structural adhesives. But this test has some drawbacks, making its results strongly dependent on the size and quality of the specimen rather than the intrinsic adhesion strength of the glue itself [6].

The aim of the present study is to propose an experimental standard procedure to evaluate the adhesion of surgical glues on soft tissue under conditions as close as possible to surgical ones. Nevertheless, it should ensure reproducibility and provide reliable quantitative values, in order to allow evaluation of the influence of some experimental parameters on adhesion. The test used for this purpose is the pressurized blister test, which has been developed for measurement of thin film adhesion on substrate [7]. This test proved to be applicable to various systems including metallic and polymer materials [7, 8]. It is appropriate to obtain a stable debonding from which the interfacial crack propagation energy is determined by a simple mechanical analysis.

EXPERIMENT

To quantitatively measure the surgical glue adhesion on soft tissue, a pressurized blister test has been chosen. Contrary to some mechanical tests already used to evaluate the adhesion of surgical adhesive [3-5] this test allows stationary crack propagation, providing a reasonably accurate and reproducible value of the interface fracture energy. This energy, also called the practical work of adhesion, is currently used in materials science to represent the bonded interface toughness.

In pressurized blister test for thin film characterization, specimen consists in a perforated substrate with a thin flexible over-coating. A fluid is injected at the interface through the perforation, thereby causing a progressive debonding. The adhesion energy is calculated from the geometry of the blister and the fluid pressure [7]. See equations 1.

$$G_c = Cph \qquad (1)$$

where G_c is the interface fracture energy, p the fluid pressure, h the blister center deflection and C a numerical dimensionless constant. The blister set-up used for the present study is equipped with a piezoelectric sensor to measure the fluid pressure. The fluid is distilled water and it is injected into the sample thanks to a micrometric screw piston or a micro-syringe. The blister geometry is recorded with a computerized fringe projection contouring profilometer. For more details on the experimental set-up see [9].

Samples preparation

Equine pericardium has been chosen as an adhesive bond layered material. This biological tissue is used in human surgery and its properties are adequate for blister testing: it is waterproof and stiff, with enough in-plane mechanical strength. Since pericardial patches are designed for cardiac and vascular repair, they can be commercially obtained and are easy to handle. Moreover, they undergo a special preparation that insures reproducibility of the testing material. For the present study we used pericardial patches from Edwards, St-Jude. They are first stabilized in glutaraldehyde to mitigate antigenicity and cytotoxicity. They are then packaged in sterile water and 2% propylene oxide that undergoes a chemical conversion to 2% propylene glycol.

Disks of 26 mm in diameter are cut in pericardial patches and dried by dabbing with paper towel. A first disk is fixed to an aluminum planar sample holder with a commercial sealant. After drying under pressure between plates, a hole is cut in the center of the disk, corresponding to the 3 mm injection hole in the center of the sample holder. A second pericardium disk is then fixed on top of the first one with the surgical adhesive of interest. Setting of the surgical adhesive is conducted according to the supplier recommendations. A thin Teflon patch of 6 mm in diameter has been installed between the two pericardium disks, on the top of the hole, to prevent surgical adhesive from flowing into the hole and make debonding initiation easier. Lastly, a thin layer of Gore-Tex® (0.1 mm thick) is adhered to the top of the second pericardium disk to provide mechanical stiffness reinforcement. The final multilayer sample obtained is schematically represented in Figure 1.

Dermabond® topical skin adhesive is tested in the present work. This adhesive contains a monomeric (2-octyl cyanoacrylate) formulation and is used to hold closed skin edges from surgical incisions.

After testing, some samples are prepared for Scanning Electron Microscopy observations. First, molding resin is injected into the hole of the sample holder in order to maintain the debonded space open. Then the whole sample on its sample holder is mould into resin before cutting and polishing to obtain cross-section surfaces for optical and SEM observations.

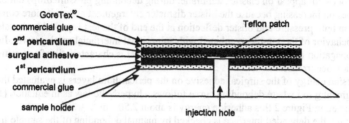

Figure 1. Schematic of the final multilayer sample prepared for blister testing.

RESULTS

Figure 2 shows typical pressure vs. height curve obtained on pericardial patches bonded with Dermabond®. The introduction of water into the hole first creates a bulge. Pressure and height increase with almost constant diameter (from a to g). When debonding begins, the lateral growth of the blister creates new volume and so leads to pressure drop (from g to j).

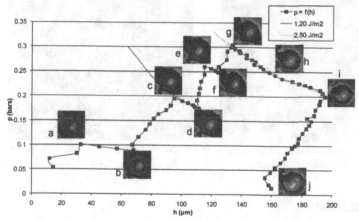

Figure 2. Typical pressure vs. height curve obtained from the blister test on pericardium/surgical adhesive/pericardium sample.

During the first step of the test, when no debonding occurs, some pressure drops are observed (c-d and e-f). Those pressure falls are not related to lateral enlargement of the bulge, as can be verified from the contour images. They are rather due to viscous relaxation when water injection

135

is interrupted. This viscous behavior is probably a combination of the constitutive viscosity of the adhesive and of the pericardium.

During the second step of the test, a more significant pressure fall is measured, linked to an enlargement of the blister debonded area (from g to j). Nevertheless, an unexpected behavior is observed on the segment i to j where the pressure decreases as well as the blister height. Such a behavior does not appear on elastic membrane: during debonding pressure drops but blister height goes on increasing because the blister diameter enlargement makes it more compliant. The decrease in both pressure and blister deflection at the end of the test is probably due also to the viscous behavior of the sample, combined with sub-critical interfacial crack growth. Indeed, crack propagation going on even at very low loading level is characteristic of sub-critical propagation occurrence.

The adhesion energy of the surgical adhesive on the pericardium layers is evaluated thanks to the curve segment g to i where debonding curve follows a hyperbolic law as in equation (1). For the curve showed in Figure 2 this adhesion energy is about 2.50 J.m^{-2}.

After testing, the debonded interface is checked by manual debonding of the sample from the sample holder and visual inspection. Some samples are also observed by SEM. It can be seen in figure 3 that the molding resin penetrates in the debonded space between the first pericardial patch (bonded to the sample holder) and the surgical adhesive. So the crack did propagate at this pericardium/adhesive interface, and the fracture energy measured with the blister test is characteristic of the surgical glue adhesion on the pericardium tissue. In this figure some small cracks are also observed between the upper pericardium tissue and the Dermabond® adhesive layer. An magnified view on one of these cracks reveals pericardium fibers network impregnated with adhesive (see Figure 4).

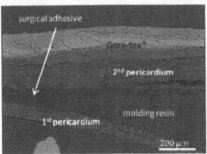

Figure 3. SEM image of the sample cross-section after debonding by blister test.

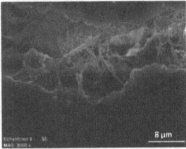

Figure 4. Enlargement of a cavity in the second pericardium from Figure 3.

Adhesion energy values

Because scattering has been observed in the adhesion energy values measured as described above, 25 samples have been tested. Moreover, the viscous relaxation dissipates some energy, leading to overestimation of the adhesion energy measured during blister testing. On the pressure vs. height curves this corresponds to a shift of the debonding segment to higher deflection values. A simple way to correct for the viscous contribution to deformation of the specimen is to graphically subtract the corresponding deflections measured on the relaxation segments during the first step of the test, before debonding initiates. Such a correction has already been used on plastic membrane [10]. After correction, the debonding hyperbolic segment is shifted towards lower height values, corresponding to lower energy levels. Those corrected adhesion energy values are used for a statistical study. It reveals that the scattering of the results is not stochastic but matches a Weibull distribution as seen in Figure 5.

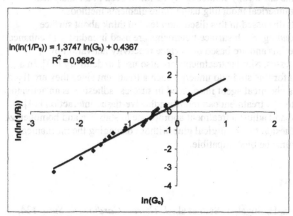

$\ln(\ln(1/P_s)) = 1,3747 \ln(G_c) + 0,4367$

$R^2 = 0,9682$

Figure 5. Weibull diagram of the safe interface probability P_S vs. crack propagation energy G_c.

DISCUSSION AND CONCLUSIONS

The pressurized blister test allowed to quantify the adhesion energy of surgical glue on soft tissue thanks to specially designed samples. Results obtained on pericardial patch/Dermabond® interface reveals very low adhesion energy values around 1 to 2 J.m^{-2}. For comparison, values usually obtained with structural adhesives on industrial materials range from tens to hundreds of J.m^{-2}. This weak adhesion has to be linked to the nature of chemical bonds between the tissue and the adhesive and to the environment this bonding occurs: blood, fat... Recent works from Artzi et al. [11] demonstrate that because of these reasons, surgical glue adhesion to soft tissue is tissue-specific.

Moreover, the surface reactivity of the tissue with the sealants seems to be inhomogeneous. Indeed, even if special care has been taken for samples reproducibility, tests reveal a large scattering in the adhesion measurements. Nevertheless, results match a Weibull distribution. This reveals a strong dependence of the debonding to the presence and the size of low adhesion areas along the interface. Such areas can indeed be observed on the debonded surfaces. This effect should be emphasized by practical surgical conditions, where tissue surface is not controlled.

Hence, mechanical behavior of the tissue also plays a role in surgical adhesive efficacy. As we observed during the test, the viscous behavior of the tissue can contribute to some energy dissipation and so enhance glue adhesion energy. Nevertheless, the mechanical behavior of soft tissues also depends on their present state. For example, aldehydes used in biomaterials - to stabilize pericardial patches or as an activator in some surgical adhesive – are known to modify pericardium mechanical properties [12]. Some improvements of the blister test analysis for surgical adhesive/tissue adhesion investigation should be obtained by taking into account the mechanical behavior of the tissue. This will be a difficult task because pericardium is not only visocelastic but also anisotropic. This anisotropy can be observed through the blister shapes. In isotropic materials, blisters are quite circular, while in the test on pericardium we observe more elliptic blisters (see Figure 2). This anisotropy cannot easily be introduced in an analytical approach but rather needs some numerical modeling using finite elements methods.

Finally, the various points discussed in this discussion lead to think about surface treatment of the tissue prior to bonding. Such surface treatments are used in industry to enhance adhesion between structural materials and are based on surface reactivity activation by mechanical and/or chemical processes. Similar treatments are also used in dentistry [13]. In a way, the pericardial patches used for this study do undergo such a treatment since they are fixed in glutaraldehyde, which is also a chemical agent introduced in surgical adhesive as an activator. Previous work also shows that storage treatment can modify adhesive-tissue interaction [14]. Then the key issue will be to find and adjust a treatment appropriate to stabilize and homogenize the tissue surface, promote its adhesion to the surgical glue without damaging the mechanical properties of the tissue and of course be biocompatible.

REFERENCES

1. M. Araki, H. Tao, N. Nakajima, H. Sugai, T. Sato et al., *J. Thorac. Cardiovasc. Surg.* **134**, 1241 (2007).
2. M. Jackson, *Am J Surg* **182**, 1S (2001).
3. K. Bresnahan, J. Howell, J. Wizorek, *Ann of Emerg Med* **11**, 2 (1995).
4. L. Ninan, J. Monahan, R. Stroshine, J. Wilker, R. Ninan et al., *Biomaterials* **24**, 4091 (2003).
5. A.J., Shapiro, R.C. Dinsmore, J.H. North, *Amer. Surg.* **67**(11), 1113 (2001).
6. R. Lacombe, *Adhesion Measurement Methods- Theory and Practice*, edited by Taylor and Francis (CRC Press 2006) pp. 409-411
7. H.M. Jensen, *Engineering Fracture Mechanics* **40**, 475 (1991).
8. R. Hohlfelder , H. Luo, J. Vlassak, C. Chidsey, W. Nix, in *Thin Films: Stresses and Mechanical Properties VI*, edited by W. W. Gerberich, H. Gao, J.-E. Sundgren, S. P. Baker (Mater. Res. Soc. Proc. **436**, Pittsburg, PA, 1997) pp.115-120.
9. B.R.M. Perrin, M. Dupeux, P. Tozzi, D. Delay, P. Gersbach, L.K. von Segesser, *Eur J Cardio-Thorac* **36**, 967 (2009)
10. C-Y. Lee, M. Dupeux, W-H.Tuan, *Mater Sci Eng* **A467**, 125 (2007).
11. N. Artzi, T. Shazly, A.B. Baker, A. Bon, E.R. Edelman, *Adv Mater* **21**, 3399 (2009).
12. R. van Noort, S.P. Yates, T.R.P. Martin, A.T. Barker, M.M. Black, *Biomaterials* **3**(1), 21 (1982).
13. J.D. Eick, A.J. Gwinnett, D.H. Pashley and S.J. Robinson, *Crit Rev Oral Biol Med* **8**, 306 (1997).
14. N.T. Samule, E. Vailhén C. Vailhé, R. Vetrecin, C. Liu, P.E. Maziarz, *J. Biomater. Sci. Polymer Edn* **19**(1), 1455 (2008).

AUTHOR INDEX

Akhtar, R., 57

Barber, Asa H., 95, 113, 119
Bidan, Cécile, 133
Boehm, C.H.J., 19
Boehm, H., 19
Braccini, Muriel L., 133
Buehler, Markus J., 89

Chen, Bin, 81
Chen, Haifeng, 125
Cruickshank, J.K., 57
Curtis, J.E., 19

de Villiers, Katherine A., 51
Derby, B., 57
Dobson, Christopher M., 33
Dooley, Clodagh, 103
Duffy, Garry, 103
Dupeux, Michel, 133

Egan, Timothy J., 51

Feng, Hailan, 125

Gao, Huajian, 81
Gardiner, N.J., 57
Gautieri, Alfonso, 89
Ge, Yanjun, 125

Hagel, V., 19
Hang, Fei, 119
Hoang, Anh N., 51

Jimenez-Palomar, Ines, 95

Katti, Dinesh R., 71
Katti, Kalpana S., 71
Knowles, Tuomas P.J., 33
Kreplak, Laurent, 25

Lee, T. Clive, 103
Li, Shaofan, 11
Lu, Dun, 113

Maitland, Duncan J., 43
Mulcahy, Lauren, 103
Mundinger, T.A., 19

Ncokazi, Kanyile K., 51

Perrin, Bertrand R.M., 133
Pradhan, Shashindra M., 71
Presbitero, Gerardo, 103

Rajagopalan, Jagannathan, 3
Redaelli, Alberto, 89

Saif, M. Taher A., 3
Sherratt, M.J., 57
Singhal, Pooja, 43
Spatz, J.P., 19
Su, Julius T., 65

Taylor, David, 103
Tisbo, Pietro, 103
Tofangchi, Alireza, 3

Vesentini, Simone, 89

Welland, Mark E., 33
White, Duncan A., 33
Wilson, Thomas S., 43
Wright, David W., 51

Yin, Yujing, 125
Yun, Song, 125

Zeng, Xiaowei, 11

SUBJECT INDEX

adhesion, 11, 19, 81, 133
adhesive, 133

biological, 3, 11, 57, 65, 89
biomaterial, 25, 33, 51, 81, 113, 125
biomedical, 71, 89, 125
bone, 71, 95, 103, 119

chemical
    composition, 43
    synthesis, 125
cluster assembly, 65
composite, 119
crystal growth, 51

elastic properties, 25, 95, 113

fatigue, 103
Fe, 51
fracture, 103

liquid crystal, 11

macromolecular structure, 65
morphology, 57

nanoscale, 71, 89
nanostructure, 81

polymer, 19, 33, 43

scanning probe microscopy (SPM),
    57, 95, 113
self-assembly, 33
sensor, 3
strength, 119
stress/strain relationship, 25
structural, 43

tissue, 133

viscoelasticity, 3, 19

Printed in the United States
By Bookmasters